*Number One: The Montague History of Oil Series*

# EARLY TEXAS OIL

The "main traveled road" in the northwest extension of the Burkburnett field. *American Petroleum Institute.*

# EARLY TEXAS OIL
# A Photographic
# History, 1866-1936

## By WALTER RUNDELL, JR.

TEXAS A&M UNIVERSITY PRESS   College Station and London

Library of Congress Cataloging in Publication Data
Rundell, Walter, 1928–
    Early Texas oil.

    (The Montague history of oil series; no. 1)
    Bibliography: p.
    Includes index.
    1.  Petroleum industry and trade—Texas—
History. I. Title. II. Series.
HD9567.T3R85   338.2′7′28209764   76-51653
ISBN 0-89096-029-01

Manufactured in the United States of America
*Fifth printing*

For Shelley, David, and Jennifer

# Contents

# Preface

BY the time the American oil industry began at Titusville, Pennsylvania, in 1859, photography was twenty years old. Not surprisingly, Colonel Edwin Drake's primitive discovery well got fully documented photographically. Since the petroleum industry in Texas developed more than three decades later, its various stages were inevitably recorded by the camera. The purpose of this book is to show the growth of the Texas oil business. Understandably, the most dramatic aspect of the business—production—attracted most photographic attention. But the other phases of the industry—refining, transportaion, and marketing—were also subjects for the camera.

As the viewer will see, the photographic record of the early Texas oil industry combines both amateur and professional efforts. Were it not for the work of such professionals as Trost of Port Arthur, Schlueter of Houston, Stephenson of Wichita Falls, and Meador of Mexia, our visual memory of the exciting days of the developing industry would be much poorer. Combined with their work was that of the snapshooter. Together they captured a visual record of the economic forces that enabled Texas to become an industrial as well as an agricultural giant. Like other kinds of historical evidence, the surviving photographs of early Texas oil do not recreate the past. Some fields were photographed extensively; others were not. Many fine pictures found their way into repositories, while others have been lost or destroyed. Although the available evidence is not uniform for all places and phases of the industry, it remains sufficiently rich and varied to enable the reader to perceive its characteristics.

Just as there can be legitimate debate about when the Texas oil industry began, so there could be dispute about when the early phase ended. For the purposes of this volume, that end came in the mid-1930's. The period covered here, 1866 to 1936, encompasses the full geographic dimensions of the industry and the colorful wildcatters who left their special mark on the business. After the huge East Texas field was developed in the early 1930's, the activities of the wildcatters became circumscribed. Corporate enterprise, usually characterized by caution, supplanted the gambling wildcatters, and the industry lost a considerable amount of excitement and color.

While the basic plan of this volume, both in the arrangement of photographs and text, is geographical, within each oil-producing region there is chronological development. Of course, the treatment of major fields is chronological, from Corsicana through the East Texas field. Production obviously has not been restricted to major fields; historically important photographs of minor fields or even lone wells are found in the epilog. Because photographers understandably emphasized the production phase of the industry, the text offers historical background to enhance appreciation of the photographers' art.

# Acknowledgments

IN collecting photographs for this book, I had the pleasure of visiting many repositories and working with unfailingly helpful curators, whose aid I happily acknowledge: Cynthia E. Keen, American Petroleum Institute; Milton Kaplan, Prints and Photographs Division, Library of Congress; G. Terry Sharrer, Museum of History and Technology, Smithsonian Institution; Mrs. Dee Brown, Corsicana Public Library; Lucile A. Boykin and Lois Hudgins, Dallas Public Library; Heddy Arlith, Fort Worth *Star-Telegram*; Joyce A. Nettles, Texas Mid-Continent Oil and Gas Association; Homer Fort, Permian Basin Petroleum Museum; R. Sylvester Dunn and Dianna Hallford, Southwest Collection, Texas Tech University; Claire Kuehn, Panhandle-Plains Historical Museum; Josephine Stayton, Wichita Falls *Times*; Chester V. Kielman, University of Texas at Austin Archives; Jean Carefoot, Texas State Archives; Joseph Coltharp, Humanities Research Center, University of Texas at Austin; Jim Pitts, Exxon Corporation; O. T. Baker, Institute of Texan Cultures; and Marion M. Branon, Houston Public Library.

The American Petroleum Institute and Texas Mid-Continent Oil and Gas Association have performed valuable services in collecting historical photographs of the industry. In most cases corporate members of these organizations contributed photographs for the files, and both repositories request that credit be given to the companies. For those companies that still exist, I have followed this request. Otherwise credit goes to the repositories.

For the sake of brevity in credits for individual photographs, I have listed only repositories. In some instances, specific collections should be cited. In the Permian Basin Petroleum Museum, the collections of Mrs. J. D. Bonner and Frank Forsyth furnished photographs. Most of the pictures from the Southwest Collection were in the C. C. Rister papers. Photographs from the Archives of the University of Texas at Austin came from the files of the Oral History of the Texas Oil Industry, the James Stephen Hogg Papers, and the Ross S. Sterling Papers. In the Humanities Research

Center, University of Texas at Austin, the Goldbeck Collection and W. S. Adkins Collection furnished photographs. At the Texas State Archives, pictures came from the W. D. Hornaday Collection. Those from the Houston Public Library are part of the Bank of the Southwest/Frank J. Schlueter Photograph Collection. The Lamar University photographs, from the Larry J. Fisher Collection, were taken by Fisher.

Many with firsthand experiences in the early days of Texas oil contributed to my understanding of the subject. Thanks go particularly to J. Chilton Lynn of San Angelo, W. A. Yeager of Midland, and Lawrence Hagy of Amarillo. Others who were most helpful include Edward Thompson of the Permian Basin Petroleum Association, Katherine L. Seewald and David Cochran of Amarillo, John F. Bookout of the Shell Corporation, Gertrude M. Teter of Lee College, and Clayton Umbach, Jr., of the Gulf Publishing Company. I have been indebted to William A. Owens, not only for his help with this project, but also for his interest in many others that go back to 1937, when our paths first crossed in Goose Creek. Dorothy Lukens typed the manuscript with characteristic cheerfulness and proficiency.

My wife, Deanna, helped with all phases of the book, especially the index. In acknowledging her assistance, I also wish to express appreciation for her unflagging interest and encouragement in all my work.

*University of Maryland*                                    WALTER RUNDELL, JR.
*June, 1976*

# EARLY TEXAS OIL

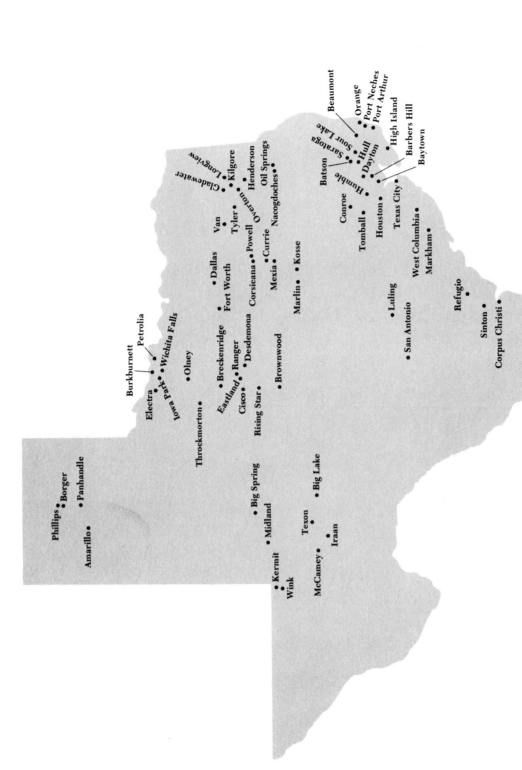

Early Texas Oil Centers

# In the Beginning

PRACTICALLY speaking, the Texas oil industry began in Corsicana in the 1890's. But the beginning of the "industry" does not connote the white man's initial awareness and use of Texas oil. The first record of that came in 1543, when remnants of the de Soto expedition washed ashore near High Island and discovered a black sticky substance they could use to caulk their boats. Prior to that and for another three centuries, Indians knew the oil springs and tar pits where petroleum in one form or another oozed from the earth. They rubbed the oozings on their bodies for medicinal purposes to relieve a variety of ailments, from skin diseases to rheumatism. A hardy few dared drink the stuff, hoping to cure digestive problems. Since this practice continued, it apparently was not lethal.

As white men settled in Texas, they discovered the surface manifestations of oil that the Indians had known, and they doubtless used the oily materials as folk medicine in much the same way as the redmen. With Texas society developing, settlers realized the need for better illuminants and lubricants, and their thoughts turned to exploiting the oil springs commercially. Although there were a few attempts to produce oil before the Civil War, extant records do not reveal any sustained efforts. Probably because of the war and its concomitant shortages, some Texans tried doggedly to make oil flow.

In the heart of the Big Thicket, Indians knew of a tar well and offered to reveal its location to white residents for fifty dollars. Not accepting the offer, some settlers located the well fortuitously in 1860 when their pigs returned from foraging with black splotches on their hides. They had wallowed in the tar well and led their owners, John F. Cotton and William Hart, back to it. Cotton decided the area had commercial possibilities and filed claim to 160 acres surrounding the tar well, which lay in what later became the Saratoga field. In 1865 Cotton signed a contract with Edward von Hartin of Galveston to drill a well. Von Hartin furnished the machinery and money for the project and became a partner in the venture.

Von Hartin hauled wagon loads of ten-foot cast-iron casings over Big Thicket trails. Over the tar well he erected a primitive tripod derrick. His drilling instrument was a bell-shaped weight attached to a rope strung through a pulley at the top of the tripod. A mule at the other end of the rope raised the weight to the top, and an automatic release sent the weight plunging into the hole. After repeating this process innumerable times, the driller reached one hundred feet, the extent of his mechanical capabilities, and the hole was abandoned. Could he have persevered, he would have struck oil, for in 1940 a good producer was drilled within ten feet of von Hartin's hole. Thus von Hartin was not destined to be remembered as the driller of Texas' first oil well. That honor would go to Lyne T. Barret.

Barret immigrated from Virginia to the Republic of Texas in 1842, a lad of ten. The family settled in Nacogdoches County, at Melrose. As he grew up, he heard talk of oil springs and tar wells in the immediate area. In 1859 he leased a 279-acre tract between Melrose and Woden but did not drill. After serving in the Confederate army, he returned to his home. Finding that oil sold for $6.59 a barrel, he resolved to follow through on his search for oil. In the autumn of 1866 he drilled on his lease and struck oil at 106 feet. The site of the well was Oil Springs, about twelve miles east of Nacogdoches.

For this pioneer producing well, Barret employed an eight-foot auger, eight inches in diameter, that he had bought in New Orleans and transported to the well site. The auger was clamped to a joint of pipe that was rotated by a drive shaft powered by steam. A tripod supported the drill, and to drill the hole deeper Barret added more joints of pipe. He withdrew the pipe by fastening a rope to the top joint, running the rope through a pulley at the top of the tripod, and having a mule draw the rope out of the well. Barret's well produced ten barrels a day. Sure of its commercial possibilities, he capped the well and traveled to Pennsylvania to enlist financial backing for the field. One speculator, John F. Carll, thought the area worth a $5,000 investment. But when he drilled an eighty-foot dry hole a mile away from Barret's well, he decided oil could be produced more cheaply in Pennsylvania and abandoned Texas. Barret's dreams of successful commercial production vanished.

Like many dreamers, Barret was merely ahead of his time. Twenty years after his failure, a crew of Pennsylvania drillers revived exploration at Oil Springs. With a cable tool rig, they put down an eight-inch casing, hitting oil at seventy feet. (Cable tool drilling operated on the percussion principle. A heavy bit, usually of the fishtail design, would be attached to the end of a cable and lowered repeatedly into the hole. As the bit went down, it had a spinning motion that helped chew into the earth. Water was run into the hole to soften the earth, and then the well was bailed out so drilling could continue.) The initial flow of between 250 and 300 barrels ran onto the ground, for J. E. Prince and his crew had not had the forethought to build tanks. When the well stopped flowing after the first day, it went on a pump. Interest in the new field ran so high that forty wells, each around one hundred feet deep, were

A tripod rig used in the Oil Springs field, twelve miles east of Nacogdoches. With a rig like this, Lyne T. Barret drilled Texas' first producing well in 1866. He struck oil at 106 feet, and the well flowed ten barrels a day. *Chief Delbert Teutsch, Nacogdoches.*

The Petroleum Prospecting Company built these two 1,000-barrel iron storage tanks at Oil Springs, the first in the state. *Chief Delbert Teutsch, Nacogdoches.*

In the late 1880's, Pennsylvania drillers revived the Oil Springs field. This superstructure protected a producing well. *Chief Delbert Teutsch, Nacogdoches.*

drilled. The Petroleum Prospecting Company diverted this flow of oil into four wooden tanks holding 250 barrels each and two 1,000-barrel iron tanks. It built a three-inch pipeline to a 2,000-barrel iron tank next to the railroad west of Nacogdoches, fourteen and one-half miles distant. This was the earliest pipeline of any length in the state.

The Oil Springs field attracted other companies, such as the Lubricating Oil Company that built and operated Texas' first refinery there. The 1890 report of the Texas Geological Survey stated that the refinery's first step was "bailing the water and the oil semi-weekly from each oil-bearing well by means of a cylindrical galvanized iron bored into a separating barrel . . . for oil. From the separating barrel the oil is drawn into carrying barrels in a wagon, which conveys it to the receiving or dump tank, into which it is emptied. This tank is located on a hillside. From the storage tanks the oil is fed by iron pipes into the oil evaporating pan, provided with a steam chest below, which heats the oil and drives off the remaining water and the small amount of naphtha. While still hot, the oil is forced by steam through a specially woven filter through a pipe into the iron shipping drum provided with wrought iron tires for rolling. The capacity of each drum is about 100 gallons. They are conveyed by wagon to the town of Nacogdoches for shipment by railway." This refining operation constituted little more than simple evaporation and filtering.

In the southeast corner of the state, serious though largely unsuccessful efforts to produce oil occurred in the 1890's. Pattillo Higgins, a resident of Beaumont, long harbored the notion that the geological formations of that area contained oil. He was intrigued with gentle elevations in the area, as well as with its gas seeps and mineral springs. Particularly, he wanted to see what lay beneath an elevation south of town, known then as Sour Spring Mound or Big Hill but later called Spindletop. Through persistent efforts, he interested several investors, who formed the Gladys City Oil, Gas and Manufacturing Company on August 24, 1892. George W. Carroll was its president and Higgins, the treasurer and manager. Higgins named the company for little Gladys Bingham, daughter of Mr. and Mrs. Ike Bingham. She received two shares of its stock.

The following year the company began drilling to determine the validity of Higgins' geological hunches. On February 17, 1893, M. B. Looney of Dallas contracted for a test well but sublet the job to Walter B. Sharp. The well was to be 2,000 feet deep, at a cost of $4.00 per foot, payable in 200-foot depths. Of the $4.00 per foot Looney received, Sharp got $2.50. Drilling began on March 22 but had reached only 418 feet in July when it was abandoned. Two years passed before the company could manage any further exploration. It contracted with Savage and Company, a group of West Virginia drillers, but their efforts went for naught. The Savage brothers' lack of success in that instance did not indicate a lack of sense for oil, for at one time they had leased most of the land in Hardin County that later had producing wells.

Twenty-five miles to the northwest of Beaumont was Sour Lake, an area noted since the 1840's for its mineral springs, redolent of hydrogen sulphide. Local pride insisted that the waters had curative powers and enough people believed it to enable a fashionable spa to operate profitably. Frederick Law Olmsted noted the healing waters in his famous 1857 journey through Texas. Whatever beneficial effects people may have experienced from bathing in the waters of Sour Lake, oil prospectors were more interested in the area because of its gas and oil seepages and sizable salt dome formation. A flurry of exploration developed in the 1890's, with the first wells dating from 1893. The Savage brothers drilled three producing wells there two years later. In 1896 Walter B. Sharp drilled a test well for the Trinity Lubricating Oil Company of Dallas. It was sufficiently successful that the company built a refinery to handle the production. This little plant turned out eight grades of oil until it burned down in 1898.

Christmas tree on a Corsicana well in the late 1890's. The term describes the valves and fittings on the top of a well to control the flow of oil. Later Christmas trees stood without the numerous guy wires. *Southwest Collection*.

# Corsicana

THE 1890's were years for oil exploration in Texas. Despite the fact that the incredibly rich deposits of Southeast Texas eluded drillers, that decade marked the beginning of successful, sustained oil production and refining in the state. Corsicana, county seat of Navarro County, was destined to raise the curtain on Texas' petroleum age.

The discovery of oil in Corsicana was hardly inevitable. It came about because the town fathers sought to shake off the decade's economic doldrums. As the center for a farming community, Corsicana prospered or faltered according to fluctuating agricultural prices. During the 1890's, those prices mostly faltered. To diversify the city's economy, a group of businessmen planned to attract industry, but the place lacked an adequate supply of water. To remedy this, the businessmen contracted for three deep artesian wells within the city.

The first of these wells was drilled in the spring of 1894 on South 12th Street, only one block from the Cotton Belt Railroad. When the well reached slightly over 1,000 feet on June 9, oil began filling the hole and rising slowly to the surface. The drillers tried to seal off the oil and save their well, but the oil continued to rise outside the casing and soak into the ground around the well. As curious onlookers gathered, one dropped a match on the ground, either carelessly after lighting his tobacco or purposely just to see what would happen. The ground was sufficiently saturated with oil to ignite immediately. After that fire was extinguished, another started when a spark from a forge where equipment was being repaired hit the ground. Thereafter, the crew dug a sump a few yards away from the well. Into that pit about 150 gallons of oil drained daily. The drillers finally got through the oil-bearing stratum and brought in their water well at 2,470 feet. The businessmen promoting the water wells had been annoyed with the delay caused by the oil and regarded it only as a nuisance.

Others, not associated with the water project, had more imagination. H. G. Damon and Ralph Beaton sent a sample of the petroleum to Oil City, Pennsylvania, for analysis. The report came back that the crude had definite commercial value. Thereupon, these two organized the Corsicana Oil Development Company and began making extensive leases near the well, offering lessors a one-tenth royalty. When the company persuaded John H. Galey to come to Corsicana to investigate its possibilities, it achieved a coup, for this noted Pennsylvania oilman was a leader in the industry. Galey thought the signs looked favorable and contracted with the company to drill on its leases. He assigned half interest in the project to his partner, James M. Guffey, and they were to bear all expenses connected with drilling five wells. In return the company would give the partners a half interest in all its current or future leases. Had the Corsicanans realized they would get the full Galey and Guffey team, their spirits would have soared, since those wildcatters had distinguished themselves with individual explorations in the 1870's of the giant field that straddled the Pennsylvania–West Virginia border and in the 1880's of the Ohio–Indiana field. As partners in 1893 they had brought in the field at Neodesha, Kansas.

Galey and Guffey set to work and drilled a well 200 feet south of the water company's artesian well. They struck oil at 1,040 feet on October 15, 1895, but the well produced only two and one-half barrels daily. A second well turned out to be a duster, but the other three, completed in May, July, and August, 1896, proved successful. These last three produced twenty-two, twenty, and twenty-five barrels a day, respectively. Such results were hardly sufficient to excite veteran wildcatters who gloried in gushers. Disappointed in this modest production and in the inability of the company to raise further capital to extend explorations, Galey and Guffey sold out for $25,000 in long-term notes. But they did not pull out of Corsicana without voicing their disappointment in eastern oil circles.

The withdrawal of eastern interests did not deter Corsicanans from further drilling. By the end of 1896, the field had produced 1,450 barrels—hardly comparing with Pennsylvania's 20 million for that year, but enough for the U.S. Geological Survey to call the increase from fifty barrels in 1895 "the greatest increase in any one state for the year." Eighteen ninety-seven saw an exponential rise in Corsicana's production—to 65,975 barrels, all within the city limits. The next year drillers moved outside the town, expanding the producing wells to 342. By the end of 1898, the Corsicana field yielded 544,620 barrels. Succeeding years saw production rise to 669,013 barrels (1899) and 836,039 barrels (1900).

As Texas' first oil field with flush production (wells flowing without the aid of pumps), Corsicana developed according to the rule of capture. In English common law, wild animals were not considered the property of any landowner but were subject to the rule of capture. That is, whoever captured an undomesticated animal on his own land was considered the lawful owner. This concept transferred to migratory underground minerals, such as petroleum. The oil belonged to anyone under whose property it lay. Therefore, lease owners in Corsicana drilled wells close together to maximize their recovery of the migratory mineral. They reasoned that if they did not capture the oil today, it might have drained into a pool beneath the neighbor's lot by tomorrow. Of course, such a practice proved extremely wasteful in terms of drilling capital and incomplete extraction of oil. But conservation of natural resources did not become a national watchword until after World War II. Following the rule of capture, Corsicana drillers sought to maximize the number of wells on a lease so that the eastern part of the town looked like a forest of eighty-five-foot derricks. This pattern of close drilling would prevail through the East Texas field of the early 1930's.

Nature aided its own exploitation at Corsicana. There oil rested in the comparatively shallow sands 900 to 1,200 feet deep. Between the oil sands and the surface lay only the black clay topsoil and strata of soft rock. Simple rotary drills could cut through to the oil with little trouble, and the wells multiplied quickly. Drillers of water wells had developed the rotary principle in the 1870's and had used it extensively around the country. The equipment consisted of a rotary or revolving platform that gripped a pipe that had a bit fastened to its end. As the hole went deeper, additional lengths of pipe were added. At Corsicana power came from a mule harnessed to the rotary platform and circling the well.

The men who introduced rotary drilling at Corsicana were the Baker brothers, M. C. and C. E., widely experienced water well drillers most recently come from the Dakota Territory. C. A. Warner, the first serious historian of Texas oil, credited the Bakers with devising rotary drilling there in 1882. To aid drilling, they poured water outside the drill pipe and the water returned, carrying rock and mud, through the pipe. Then they decided the process would be more efficient if reversed, with water going through the pipe. Even so, this circulation of water and cuttings proved slow and tedious until the Bakers realized they could harness power from a windmill to force water through the pipe. When windmills were not handy, they used steam pumps. The Bakers arrived in Corsicana with their mechanical know-how in 1895, hoping to participate in a boom. Their rotary drills were so successful in striking oil that by mid-1898 all drillers in the field had converted from cable tool rigs. And little wonder. A rotary rig could drill a 1,000-foot well in thirty-six hours at an average cost of $600. Among those immediately attracted to the Bakers' striking innovation were owners of the American Well and Prospecting Company, the firm that had dug the 1894 water well. These men, Charles Rittersbacher, H. G. Johnston, and Elmer Akins, entered into an agreement with the Bakers to improve and manufacture rotary equipment. The company flourished and became a leading supplier for the industry in the Southwest.

Despite Corsicana's obvious success as an oil field, local businessmen expressed concern over the long-range impact of oil on the economics of the community. They understood that mere production of petroleum was not the desired goal, and yet they had not been successful in creating a sufficient market for their oil to generate the desired financial profit. They thought out-of-state capital would be eager to invest in developing the field and marketing the product, but found that most developers wanted the locals to put up unreasonable amounts of cash. In other words, the outside interests usually wanted to skim the profit off the top while the locals bore most of the risk. Hoping to attract the interest of an experienced oilman from the East, Mayor James E. Whiteselle wrote Joseph S. Cullinan, inviting him to visit Corsicana. Cullinan, who had come up through the ranks of a Standard Oil affiliate in Pennsylvania, decided to have a look at Corsicana's prospects in October, 1897. He liked what he saw and determined to transfer his managerial talents to Texas.

Cullinan knew if respectable profits were to be made from Corsicana's oil, the ultimate goal was successful selling. Before that could be reached, Corsicana needed storage tanks, pipelines from wells to the tanks, and ultimately a refinery. Luckily, he could construct the first two with his own company, Petroleum Iron Works, head-quartered in his hometown, Washington, Pennsylvania. For the refinery, he would have to solicit outside capital. This was not hard to find. Former associates in Stand-ard Oil, Calvin N. Payne and Henry Clay Folger, agreed to finance its construction. The partnership would be known as J. S. Cullinan and Company, with Payne and Folger furnishing all the capital—$50,000 each. Cullinan contributed management of the operation and received a $5,000 annual salary.

In June, 1898, Cullinan began construction of his refinery on a 136-acre tract in southwestern Corsicana. He hired E. R. Brown to supervise building the plant. By December the installation was complete—all four stills, each twelve feet in diameter and thirty feet long, arranged in a row. Each boilerplate still held five hundred bar-rels. On Christmas Day W. C. Ralston fired the first still, and refining a "batch" of crude oil began.

Unlike the primitive evaporation process used in the earlier Oil Springs re-finery, this batch refining was a bit more complex. Fittingly, the crude in the stills was heated by oil burning in the furnaces. As the temperature in the stills rose, various fractions of the oil vaporized separately. As gases collected at the top of the still in a large dome, they were piped through coils into a condensing tank. There the vapors liquified with water cooling. These distillates received further refining to re-move impurities. Two vertical agitator tanks, measuring over sixty feet, removed most of the color, which made the kerosene—by far the major petroleum product—more appealing to the buyer. At the acidizing plant, the refined oil was treated with sulphuric acid to remove most of the sulphur from the oil. From the earliest days of the Texas oil industry, the sulphur content of petroleum has plagued refiners. High sulphur content in kerosene resulted in a smoky, malodorous lamp. In motor oils and gasoline, it has meant fouled engines and noxious emissions. For those who worked in or lived near refineries, the acidizing plants with their penetrating aroma of hydro-gen sulphide (akin to, but worse than, rotten eggs) have lent body to prevailing breezes.

With the refinery operating regularly, the Corsicana business community exulted. When the first tank cars of refined products were ready for shipment, the city organized a celebration and decked one tanker with a banner proclaiming "Cor-sicana Petroleum Oil. Greatest Discovery Ever Made in Texas. Quality as Good as Pennsylvania's. Quantity, Unlimited." Among those boosters photographed on that occasion was Claude Jester, a relative of Beauford Jester, governor from 1947 to 1949.

To maintain the prosperity of Corsicana's oil community, Cullinan had to ensure a steady market for its products, and he set out systematically to do so. He persuaded various businesses, such as the Waco Ice Company, to switch from coal to oil by demonstrating that they got better results at a lower price. To his younger brother Michael he delegated the responsibility for convincing railroads to convert locomotives from coal to oil burners. Despite a successful demonstration on a passenger run between Corsicana and Hillsboro, forty miles away, the Cotton Belt Railroad did not adopt the new fuel. In 1901, however, the Houston and Texas Central Railroad, also running through Corsicana, did. Cullinan pointed out to the city of Corsicana that in areas with any oil seepage, as around wells and storage tanks, there was little dust and that the city could control dust on streets by oiling them. Thus Corsicana became the first place in Texas with dustless streets. Soon other cities bought Corsicana oil for the same purpose: Fort Worth, Honey Grove, Waco, and Greenville. Through Cullinan's efforts, another first for Corsicana was natural gas for illumination.

With Corsicana's refining beginning at the turn of the century, it spanned a transitional period in the American oil industry. Refining made its debut there at the end of the illumination period, 1859–1900, when the chief petroleum product was kerosene or coal oil. Then it participated in the fuel oil decade, 1901–1910, when refined petroleum began to replace coal in the furnaces of locomotives and ships. Cullinan's seeking to sell his products to railroads typified this decade. When motor cars came to the fore, the industry entered the gasoline era.

Through Cullinan's efforts, Corsicana had established itself as the premier city of the Texas oil industry. It demonstrated the possibilities of flush production, commercial refining, and successful marketing. In the process it had also demonstrated a reckless waste of natural resources. The Texas legislature noted this dissipation and passed a regulatory law, hoping to prevent it. On March 29, 1899, "An Act to Regulate Drilling, Operation and Abandonment of Petroleum Oil, Natural Gas and Mineral Water Wells, and to Prevent Certain Abuses Connected Therewith" became effective. It provided for encasing wells, plugging abandoned wells, and capping gas wells until they were piped for commercial usage, and limited the burning of gas for illumination between 8 A.M. and 5 P.M. Violation of any provision entailed a $100 fine. This act formed the basis for future efforts of the state of Texas to regulate wasteful production to conserve irreplaceable natural resources.

The successful emergence of Corsicana as an oil center depended on Joseph Cullinan. The area had provided an excellent training ground for his further career in Texas oil. After he shifted his interest to Southeast Texas, he sold the Corsicana refinery to Magnolia Petroleum Company. The field continues to produce from deeper levels. In the 1920's Powell, just a few miles east of the city, became the site of important discoveries. In 1941 Magnolia abandoned Texas' first commercially successful refinery, having consolidated operations at its Beaumont plant. Historically conscious, Magnolia's successor, Mobil Oil, has kept one of the four original stills as a monument to the origins of the Texas oil industry.

The rotary rig that drilled Corsicana's first well in 1894. Now in the Smithsonian's Museum of History and Technology. *Smithsonian Institution*.

*Early Gripping Device*

*Left*: This well in the front yard of C. D. Speed, 912 East 8th Avenue, Corsicana, began producing September 23, 1898. *UT Archives*. *Right*: C. E. (pictured) and M. C. Baker developed the rotary drill for water wells in the Dakota Territory in 1882. They were among the first drillers in Corsicana. This gripping device kept the drill pipe from slipping as it turned. *Smithsonian Institution*.

An early portable rotary rig, one stage beyond that employed in the first Corsicana well. *Smithsonian Institution*.

Founders of Corsicana's refinery, 1898, the first significant one in Texas. *Left to right*: J. S. Cullinan, Craig Cullinan (boy), George Smith, F. T. Whitehill, Dr. Michael P. Cullinan, Willis C. Collier, W. H. Page, John Cullinan (boy), W. C. Proctor, E. R. Brown (who built the refinery and was later president of Magnolia Petroleum Company), Danny Burk, W. T. Cushing (who constructed Texas' first pipeline), Frank Cullinan, Ed Wright, and Frank Perfield. *American Petroleum Institute*.

E. R. Brown (*left*) and W. H. Page. Brown built the Corsicana refinery in 1898. *Texas Mid-Continent Oil & Gas Association*.

*Left*: Corsicana officials invited J. S. Cullinan, well-known Pennsylvania oilman, to investigate the commercial possibilities of their oil in 1897. He immediately appreciated them and organized his company, a predecessor of the Magnolia Petroleum Company. *Texas Mid-Continent Oil & Gas Association.* *Right*: Henry Clay Folger, a partner in the Cullinan Company, agreed to finance the construction of the Corsicana refinery. He later endowed the magnificent Folger Shakespeare Library in Washington, D.C. *Texas Mid-Continent Oil & Gas Association.*

*Left*: On December 25, 1898, W. C. Ralston fired the first still at the Corsicana refinery, which he helped build. He re-enacts that historic event in this photograph. *Texas Mid-Continent Oil & Gas Association.* *Right*: A close-up of the drive chains on a rotary rig. *Exxon Corporation.*

The Corsicana refinery under construction in 1898. It operated until April 1, 1941. *American Petroleum Institute*.

By 1900 derricks dominated residential Corsicana. *Texas Mid-Continent Oil & Gas Association*.

In the nineteenth century, as well as twentieth, Texans dealt in superlatives. Among these tank car boosters were Ralph Beaton, Sr.; Captain J. E. Whiteselle, then mayor of Corsicana; H. G. Damon; Captain A. Angus; H. T. McCallan; Claude Jester; and H. L. Scales. *Texas Mid-Continent Oil & Gas Association*.

In 1923 production began at Powell, eight miles east of Corsicana. A rotary rig, such as the above, hit oil at around 3,000 feet. *American Petroleum Institute*.

On January 10, 1901, Captain Anthony F. Lucas brought in the world's greatest gusher at Spindletop. He hit oil at 1,020 feet; this well flowed between 75,000 and 100,000 barrels a day. The field ushered in a new era that transformed the American petroleum industry. *American Petroleum Institute*.

# Spindletop: Beaumont and Environs

JANUARY 10, 1901—a date that revolutionized the petroleum world! Since 1892 Pattillo Higgins had promoted the possibilities of oil around the salt dome known as Sour Spring Mound, Big Hill, and later Spindletop. Drillers, such as Walter B. Sharp, had put down test wells but with disappointing results. Higgins and the Gladys City Oil, Gas and Manufacturing Company did not despair but kept searching for a driller who could find the oil trapped by the salt dome. Higgins' advertisement in a trade journal attracted the attention of Captain Anthony F. Lucas, a mining engineer who had experience with Louisiana salt domes. After corresponding with Higgins, Lucas journeyed to Beaumont to look at the land. On June 20, 1899, the company sold its leases to Lucas for $33,150. Recognizing Higgins' assistance, Lucas awarded him a 10 percent interest in the lease. A native of Austria, Lucas had studied engineering at the Polytechnic Institute at Graz. After service in the Austrian navy, he had visited his uncle in Michigan and determined to seek his fortune in America. Becoming naturalized in 1885, he prospected for gold in Colorado before mining salt in Louisiana. Although not an oilman, Lucas brought to his task the training of an engineer and long experience with salt domes.

Lucas' first well went to 575 feet and produced only a small amount of oil before his money ran out. Local financiers scoffed at his efforts and offered no assistance. Upon the advice of Dr. William B. Phillips, director of the Texas Geological Survey, Lucas sought the support of Galey and Guffey, the disillusioned Corsicana veterans who nevertheless maintained an interest in oil possibilities along the Gulf coast. In their Pittsburgh office Galey and Guffey became convinced that Lucas was a good bet and agreed to finance him. They would assume the financial liabilities of the Gladys City Company in exchange for a sizable percentage of its holdings.

The backing of Galey and Guffey laid the foundation for Lucas' success at Spindletop. Galey arranged with the Hamill brothers—Curt, Jim, and Al—who had drilled for him in Corsicana, to go to Beaumont as drillers for Lucas, with Galey and

Guffey guaranteeing payment and supplying all the pipe. The Hamills would get $2 per foot for a 1,200-foot well. They spudded in, or began, the well on October 27, 1900, with a rotary rig transported from Corsicana. There they had used water to soften the ground, cool the bit, and flush out the cuttings, but at Spindletop water could not accomplish the last objective. Curt Hamill improvised an ingenious expedient. He drove some cattle into a shallow pond nearby, hoping their milling around would produce mud that would be sufficiently viscous when pumped down the drill stem to bring up the cuttings. The experiment worked, and from that time forward drilling mud became a mainstay in rotary operations.

As with many technological innovations, mud possessed some highly desirable attributes other than that originally intended. It came to have four main functions. Most important, it counteracted pressures exerted by gas, oil, or water, preventing dangerous and costly blowouts. It served as a lubricant, cooling the whirling bit and keeping it from burning out from the intense friction. It flushed out cuttings from around the bit, eliminating frequent stops to clean away fragments of sand, clay, and rock. And it plastered and exerted pressure on the sides of the hole, thereby lessening the danger of cave-ins and helping prevent water or gas from entering through the side walls. When the industry became more advanced and technology more sophisticated, artificial muds were devised to meet exacting specifications. Most wells struck the kinds of shales and clay needed to provide the basic ingredients of mud, and to these chemicals were added. If natural muds were not available, drillers could import dehydrated mud in paper bags! To these they added the appropriate parts of water, chemicals, and weighting agents.

The Hamills encountered numerous difficulties with their well. For their boiler they could get only water-soaked logs, so the fireman had difficulty keeping enough fire going to produce steam. The quicksand through which they had to drill slowed the operation to a few feet per day. One of the four members of the crew, Henry McLeod, left because the work was too strenuous. The remaining members, Curt and Al Hamill (Jim had gone back to Corsicana) and Peck Byrd, had to go to eighteen-hour shifts or tours (pronounced towers) so that drilling could continue. By Christmas Eve, Galey, who had come from Pittsburgh to observe the test well, realized that the crew was exhausted and recommended that they shut down for a Christmas holiday, the well having reached 880 feet.

The Hamills returned to Corsicana to spend the vacation, arriving back at Spindletop on New Year's Day. In a week they had drilled to 1,020 feet, where the bit got deflected by a crevice. Needing a different kind of bit, Al wired Jim in Corsicana to send a fishtail bit, which arrived on the train on January 10. Al inserted the new bit to a depth of 700 feet, when the well started regurgitating mud high up the derrick. Al noticed that it was 10:30 A.M. Immediately after the mud, the four-inch drill pipe shot up through the derrick, knocking down the crown block. As the pipe ascended, several joints broke off at a time. Then the well became quiescent. The disheartened drillers surveyed their spoiled well and started shoveling the muck off the derrick floor. Suddenly, the well disgorged vast quantities of mud with an explosive roar. The frightful noise continued as gas propelled upward. Then oil shot above the derrick too—the black plume doubling the derrick height, more oil gushing than any Texan had seen. More than anyone had ever seen, for the Lucas discovery well at Spindletop broke the world's record. In its first nine days, the well produced 800,000 barrels of oil. But what to do with that oil flowing freely over the rice fields? Lucas hastily commandeered forty four-horse teams to throw up levees to contain the oil. They also plowed under the oil-soaked grass to lessen chances of fire. On the third day, however, a careless smoker set the grass afire, but the fire was smothered before it got to the well.

Lucas' ultimate concern, of course, was how to regulate the well. As he put it to the drillers, "Now that we have got her, boys, how are we going to close her up?" Since the gusher spewed up large rocks periodically, the crew hesitated to cap the well too soon for fear that a rock would knock off the valve. On January 16 the Hamills imposed their fittings over the geyser of oil: an eight-inch T that had an eight-inch gate valve and an eight-inch nipple screwed into it. The side opening in the T connected with a six-inch pipe that had a gate valve through which the oil flowed. The Hamills bolted this apparatus over the casing until Al could prepare the drill pipe to receive the T. By the tenth day it appeared safe to try this daring feat. With Galey and J. S. Cullinan offering advice, Al undertook the job of sawing through the iron pipe to remove a "protector" that could not be unscrewed and then dressing the threads so that the T could be screwed on. As he worked in a deluge of oil, Hamill knew that one spark from his hacksaw would blow him—and the Lucas well—into kingdom come. He worked steadily for several hours, finally screwing the T onto the casing, and his brother Curt closed the valve. "Just like that, it was over," Al commented.

Only the gushing of the Lucas well was over, for the boom had just begun for Beaumont. Within three months the sedate city of 9,000 had swelled to 50,000 as speculators, oil-field workers, adventurers, and spectators swamped the town. The first flurry of visitors included oilmen who wanted to see for themselves what had happened at Spindletop. J. S. Cullinan had hurried down from Corsicana, bringing with him a noted visitor, Mayor Samuel M. ("Golden Rule") Jones of Toledo, Ohio. Jones was a prominent manufacturer of oil equipment and an outstanding political reformer. William S. Farish, fresh out of law school in Mississippi, and Robert L. Blaffer of New Orleans hurried to the scene and decided to seek their fortunes in Texas oil. Later they were among the first officers of the Humble Oil and Refining Company. Former Governor James S. Hogg and his partner James W. Swayne formed an important syndicate, selling Spindletop leases. James Roche also became amazingly adept at dealing in the complicated maze of oil leases. From the Chicago area John W. ("Bet-a-Million") Gates arrived to survey the prospects—and decided they were worthy.

As could be imagined, speculation was rife throughout Beaumont and the surrounding countryside. Pig wallows sold for $35,000 and cow pastures for $100,000. Land 150 miles from Beaumont sold for $1,000 per acre, and land within the proven Spindletop field sold for $900,000 an acre. Because of the sharp and shady practices of many lease dealers, the area got the nickname of Swindletop. The development of the field was likewise frenzied, with wells being drilled as close together as physically possible. In fact, the desire to exploit the field quickly ran so high that on occasion four wells were drilled beneath one derrick floor.

Spindletop soon learned the cost of crowded drilling. On a morning in September, 1902, a driller thoughtlessly pitched his cigar off the derrick floor, starting a tragic fire in the Hogg-Swayne tract. As the flames spread from wells and earthen reservoirs through the field, it became manifest that only an organized firefighting effort could save the field. Oil company leaders and Beaumont officials appealed to J. S. Cullinan to head the team. He agreed, exacting from them the right to enforce his orders at gunpoint. For a week Cullinan and his team worked arduously without sleep, combatting the searing flames with sand and steam. With the battle won, Cullinan had to be hospitalized, completely exhausted, eyes and lungs the worse because of constant gas fumes.

From Spindletop sprang two giants of the modern industry, the Texas Company and Gulf Oil. J. S. Cullinan had joined with John W. Gates and Arnold Schlaet to form the Texas Fuel Company, which soon dropped "Fuel" from the corporate name. They built their refinery at Port Arthur, where ocean-going tankers could dock. The 1902 Spindletop fire had threatened to wipe out the young company, but Cullinan had saved its tremendous stake in the field at grave personal risk. Gates and Schlaet, the financial backers, agreed that Cullinan's risk had been too great, since he was irreplaceable in the company. The Mellons of Pittsburgh backed the J. M. Guffey Petroleum Company (actually a Guffey-Galey-Lucas partnership) that would soon become Gulf Oil. Guffey piped oil to Port Arthur, where, before 1901 was out, he had erected a refinery. Another established Pennsylvania oilman, J. Edgar Pew, came to Beaumont and invested heavily in production for his Sun Oil Company. He had, in fact, built a refinery at Marcus Hook, Pennsylvania, to refine Spindletop oil even before he purchased the oil!

Accompanying the feverish drilling at Spindletop came construction of storage, refining, and transportation facilities. When railroads seemed loath to furnish tank cars to haul out Beaumont crude, former Governor Hogg, influential in establishing the Texas Railroad Commission, brought pressure and secured the needed cars. Beaumont's teeming population had to be housed somehow. Many workers slept in tents, paying fifty cents a night, with no charges for hordes of malaria-bearing mosquitoes. Ramshackle huts sprang up, and the settlement of Gladys City, adjacent to

This Chapman rotary rig resembles that used by Captain Anthony F. Lucas to drill the Spindletop gusher. Although long thought to be a photograph of the Lucas well, this picture probably dates from a later drilling. No contemporary photographs of the Lucas well show a building in the background. The inscription most likely meant it was the driller's (*arrow*) first well. *American Petroleum Institute*.

Spindletop, overflowed with roughnecks. Saloons, whorehouses, and gambling parlors flourished, ready to relieve the farm boys who flocked to the oil fields of their hard-earned wages.

Much of the excitement of Spindletop was captured by photographers: the mushrooming of derricks in the field, the crush of folks at the railroad depot, and the ruinous fires that swept the field. One bit of color unfortunately escaped the camera, but William A. Owens in his interview with Burt E. Hull, an oil pioneer, captured it. Jim Hogg, the portly former governor, paid close attention to his Spindletop interests. When in Beaumont he stayed at the Oaks Hotel on Calder Avenue. On one occasion a thunderstorm flooded Beaumont, a frequent occurrence. Hogg needed to visit the Texas Company's office downtown on Pearl Street, and the flood was not about to deter him. He hired a Negro to pull him in a skiff. From the hotel Hogg took a rocking chair to put in the flat-bottomed skiff and settled his three hundred pounds into it, first pulling his trousers up over his knees. To keep the hot morning sun off, he hoisted a huge red and white umbrella. Like Cleopatra on her barge, Hogg floated placidly downtown while everyone else waded on the sidewalks, gawking in amazement and some amusement.

*Left*: As drilling continued at Spindletop, these kinds of equipment—rotary, swivel, and elevators—were employed. *Smithsonian Institution*. *Right*: After Lucas struck oil at Spindletop on January 10, 1901, Beaumont grew from 9,000 to 50,000. This scene at the railroad station documents the way the population zoomed. *American Petroleum Institute*.

The drilling crew angled this 1901 Spindletop gusher to the side. Note the well-dressed women near the derrick. *Sun Oil Co*.

Since the derrick had been removed from this 1901 Spindletop well, the operator must have allowed it to gush for exhibition purposes. *Library of Congress*.

The Texas Flora Oil Company's gusher caught fire shortly after midnight on September 12, 1902, the first well to burn at Spindletop. *Sun Oil Co.*

On September 13, 1902, people rode out from Beaumont in surreys to observe the fire begun at the Flora gusher. *Texas Co.*

The Flora fire destroyed about twenty wells, many storage tanks, and a pumping station. One well-dressed woman turned out for the occasion (*far right*). *Texas Co.*

Spindletop's Boiler Avenue, 1903. The densest drilling in Texas. Note the different derrick heights. *American Petroleum Institute.*

In 1902 the Port Arthur photographer F. J. Trost captured this calm scene at Spindletop. Someone later obliterated Trost's name from the lower right corner of this print. *Gulf Oil Co.*

Aftermath of the 1903 fire in the Hogg-Swayne tract at Spindletop. Former governor James S. Hogg and James W. Swayne developed a large portion of the field. *Exxon Corporation.*

Another view of the 1903 fire in the Hogg-Swayne tract at Spindletop. *Exxon Corporation.*

James S. Hogg in 1905, when Spindletop's production continued to amaze the world. Unidentified man on right. *UT Archives.*

Four teams moved this boiler on a Sun Oil lease at Spindletop. The teamsters are (*left to right*): Charles Radka, Jim Myers, Dill Hill, and Al Castom, the driver. Those with arms akimbo and the mounted man in the rear are unidentified. *American Petroleum Institute*.

Guffey Petroleum Company roughnecks at Spindletop, 1903. Names not certain. *Front row, left to right*: J. Whittier, Shader, C. Burns, John Henderson, Snider, J. Sivily, T. Potter, and Sam Weaver. *Top*: Haskell Perkins, G. Barham, D. Stevens, Roy Reader, J. Leslie, Scotty Long, and J. Long. *American Petroleum Institute*.

This open sump at Spindletop, 1904, held 3 million barrels of oil. *American Petroleum Institute.*

Spindletop workers could come to Beaumont to sleep in style for fifty cents a night. *Sun Oil Co.*

Those who wanted to live closer to the job could stay at Mrs. Keenan's boarding house in Gladys City, the town that mushroomed next to Spindletop. Only Mrs. Keenan, standing on the porch, can be positively identified. Others in the picture include Thompson, McGintry, J. Whittaker, W. T. McGussey, C. Erwin, McChaffee, H. B. Simcox, Harry Bell, C. A. Brown, T. Pastur, Shader, Roy Reeder, Charles Keenan, and Jim Keenan. *Southwest Collection.*

Gladys City, Spindletop. Named for Gladys Bingham, minor daughter of Mr. and Mrs. Ike Bingham. The Binghams were stockholders in the Gladys City Oil, Gas and Manufacturing Company, organized in 1892 to exploit the salt dome known as Spindletop Hill. Pattillo Higgins was another shareholder. *Texas Mid-Continent Oil & Gas Association.*

The ball penetrated the fellow's wrist and his revolver fell to the ground. As Harry covered the second man, who threw up his hands, Old King Brady kicked the barrel away, leaving the badman hanging in the window, unable to move.

Spindletop quickly entered popular fiction. *Ethyl Corp.*

*Left*: So that heavy machinery could be hauled through deep mud, plank roads like this one near Beaumont were laid. *UT Archives*. *Right*: Longe #2, brought in by wildcatter Glenn H. McCarthy near Beaumont on April 4, 1936. Two 30-foot joints of 7-inch casing blew out as the well began gushing. The well caught fire twice, this picture showing the second. McCarthy gave up the well as a total loss. *National Archives*.

At the fortieth anniversary of the Spindletop discovery, the drillers observe the monument commemorating the Lucas well. *Left to right*: Al W., Jim G., and Curt G. Hamill. On this historic well, the Hamills pioneered in using mud to control gas pressure and in capping the well with an 8-inch T. *Texas Mid-Continent Oil & Gas Association*.

Horse-drawn welding unit on the 8-inch El Dorado–Beaumont pipeline in 1926. *American Petroleum Institute*.

Drilling in the Nome field, twenty miles west of Beaumont. The farmer leasing his land hoped for more success with the oil well than he had with the water well in the foreground. A high wind sweeping the coastal plain snapped the neck of the windmill. *Smithsonian Institution*.

Spindletop: Beaumont and Environs / 51

From these derrick tops at High Island (forty miles southwest of Beaumont), one could see the Gulf of Mexico. June, 1934. *Humanities Research Center, UT*.

Before the sides of a storage tank were erected at a Beaumont refinery in 1901, the bull riveter fastened steel plates horizontally. *Library of Congress*.

After the first tier of tank siding was in place, the bull riveter worked vertically. *Library of Congress.*

A pumping station and loading terminal at the J. M. Guffey refinery at Port Arthur in 1901. Pipelines, pumping stations, and railroad tank cars formed the transportation network to move crude and refined oil to refineries and consumers. *Library of Congress.*

*Left*: Storage tank of the Lone Star and Crescent Oil Company, Beaumont, 1901. The company was bought by Sun Oil in 1904. Sow and shoats appear unconcerned with the nearby industrial activity. *Library of Congress. Right*: Beaumont's Magnolia refinery, taken from the air across the Neches River in the mid-1930's. *National Archives.*

*Left*: John H. Galey, partner in the firm of Guffey and Galey of Pittsburgh, Pennsylvania, at a well site. These partners helped develop Corsicana and later Spindletop. Galey aided Captain Lucas and the Hamill brothers (drillers) in capping the discovery well at Spindletop after nine days of gushing. *Southwest Collection. Right:* James M. Guffey, Eastern capitalist who helped develop the early Texas oil industry. *Texas Mid-Continent Oil & Gas Association.*

Headquarters for trading and buying Spindletop leases, 1901. *Library of Congress.*

For those who didn't want to trade at the Oil Exchange Building, the curb offered ample opportunity. *Shell Oil Co.*

The Sun Oil Company opened its first Texas office in this Beaumont building in 1901. *Left to right*: Sun employees Frank Maxwell, Wes Sturm, unidentified, and Bill Sturm. Sun was so encouraged by the prospects that it built a refinery for its Texas oil at Marcus Hook, Pennsylvania, several months before it acquired any Texas oil property in 1902. *Sun Oil Co.*

The Texas Company developed from the Spindletop field. This was its first headquarters. *Texas Co.*

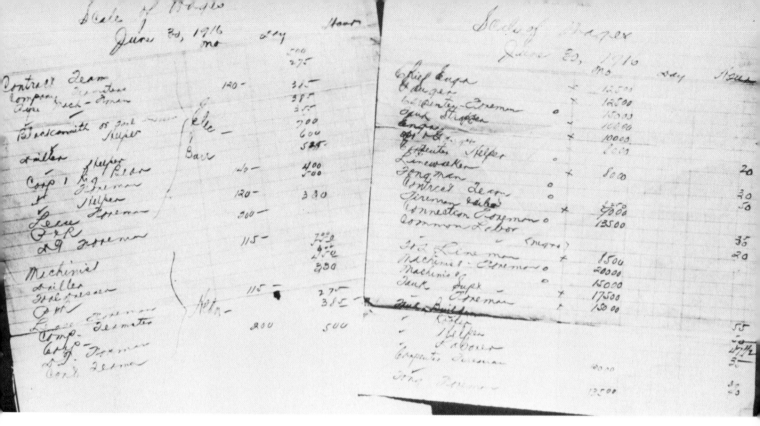

Wage scales for production (*left*) and pipeline (*right*) employees of Beaumont's Magnolia Petroleum Company in 1916. *American Petroleum Institute*.

IN 1901

IN 1935
(Both Views from Same Location)

Port Arthur, seventeen miles southeast of Beaumont on Sabine Lake, has immediate access to the Gulf. Because of its proximity to Beaumont, it developed into a natural refining and transportation center. It boasts two great refineries, the Gulf and Texaco. This photograph shows the growth of the Gulf refinery over three decades. In the earlier picture cattle placidly file by the distilling unit. *Gulf Oil Co.*

Spindletop: Beaumont and Environs / 57

Asphalt, the heaviest product of oil refining, used to pave a Port Arthur street in 1904. *Texas Co.*

Crude stills at the Texas Company refinery in Port Arthur, 1911. *Texas Co.*

A fire at Port Arthur's Texas Company storage terminal, June 26, 1911. *Texas Co.*

The Texas Company refinery at Port Arthur at night, 1913. *Texas Co.*

Spindletop: Beaumont and Environs / 59

The first steel oil tanker, the S.S. *Standard*, was built in 1888. *American Petroleum Institute*.

Texas oil begins to find distant markets under sail. This photograph by Trost shows the first vessel loaded with export oil at the Port Arthur docks, May 17, 1901. *UT Archives*.

Disaster struck the *City of Everett* at the Port Arthur dock in 1903. The explosion blew Captain Fencon off the vessel. *American Petroleum Institute*.

The *Delaware Sun*, a six-masted schooner, was converted to transport oil in 1912. It had a capacity for 37,400 barrels. This vessel loaded crude oil from Spindletop at the Sun Oil Company's Port Arthur docks. Oil was carried under sail through World War I. *Sun Oil Co.*

Spindletop: Beaumont and Environs / 61

*Left*: The Texas Company island and wharves in Port Arthur, 1923. Note Intercoastal Waterway at top of picture. *National Archives*. *Right*: Tankers loading at the Texas Company docks in Port Arthur, June 25, 1930. *Southwest Collection*.

The Gulf refinery on Taylor Bayou, Port Arthur, mid-1930's. *National Archives*.

Drilling at Port Neches, on the south bank of the Neches River now contiguous with Port Arthur. Back side of the drawworks on Polk #2. Engine is a 10″ x 10″ Union Tool Company twin-cylinder drilling machine. *UT Archives.*

Sour Lake, Texas, *May 20 1897*
(HARDIN COUNTY.)

Dear Since writing you today I have received telegram awarding me the contract at Galveston also letter from the H & T. C R R Co wanting me to take there contract to move dirt off of ther road, also have struck more ail and the prospect is getting so much better I hate to leave here. I just hardly know what to do for the best

Mr Brown also wired me to come to Galveston I answard that had struck ail and could not untill Sunday as Monday at earlist

Write me dear.

Walter

And lots of love

Among those who stayed at the Sour Lake Hotel was Walter B. Sharp, as this letter of May 20, 1897, indicates. He drilled his first successful well there in 1896. *Dudley Sharp, Houston.*

# Saratoga, Sour Lake, and Batson

ONCE Spindletop had validated Pattillo Higgins' salt dome theory, prospectors began searching for similar formations in Southeast Texas. In close proximity were three communities with salt domes covering vast lakes of oil: Saratoga, Sour Lake, and Batson. As early as 1860 John F. Cotton had identified a tar seepage near Saratoga, but his 100-foot well produced nothing. Then in the mid-1890's a bit of the Sour Lake potential was tapped, but operations were not sustained there. Only after Spindletop would serious investigation begin. At Saratoga in the fall of 1901, the discovery well, Hooks #1, came in yet yielded little. Only with Hooks #2, gushing on March 13, 1902, did Saratoga establish itself as a major field. A Galveston newspaper reported the news quaintly: "About ten o'clock this morning a telephone message from Saratoga brought information that the Hooks well No. 2 was performing in a way that would equal any of the freaks on Spindle Top." Hooks #2 yielded five hundred barrels daily, nothing to compare with Lucas' "freak" well, but highly respectable. Another fine well a year later set off a rash of further drilling, and the Texas and Gulf companies linked the field with pipelines to Beaumont.

The first positive indication that Sour Lake had vast potential came on March 7, 1902, when the Atlantic and Pacific Company struck oil at 646 feet. The output of that well amounted to over 30,000 barrels a day. Other wells drilled at Sour Lake in 1902 varied in production between 500 and 10,000 barrels daily, and Sour Lake had qualified as a major field.

J. S. Cullinan's Texas Company made Sour Lake a cornerstone of its production. Cullinan hired Walter B. Sharp as a master driller, and Sharp returned to his former haunts in 1902. He put down three wells near the resort hotel but wanted to keep their potential hidden from competitors as long as possible. The first two he delayed completing. At 780 feet the third gave every sign of being a gusher, so Sharp gambled on the weather. Early in 1903, on a night with lowering clouds, Sharp brought in his gusher, Fee #3, which drenched surrounding trees. As Sharp gambled, the skies then opened and washed from the trees all evidence of oil. Visitors were fooled, but the Texas Company took up options to sell Sour Lake production at sixty cents a barrel. From that well's first month, the company netted $150,000, since the cost fell to ten cents per barrel.

The main significance of the Texas Company's venture at Sour Lake was that by purchasing most of the productive tract, it ensured for the young company an ample supply of crude. Unfortunately, drilling there followed the same pattern as at Spindletop, with wells jammed up against each other. Sour Lake's "Shoestring" district, as the name indicates, was a narrow lane with intense drilling. When the field peaked with flush production in September, 1903, it yielded between 50,000 and 60,000 barrels daily. By the end of that year, with the depletion of gas pressures, the flow declined and half the 150 wells were abandoned. In those early days, drillers found oil at depths ranging between 750 and 1,050 feet. With deeper drilling in 1916, 1923, and 1935, the field once again became a significant producer.

A half-dozen miles southwest of Saratoga, Batson soon joined the other salt dome areas as a major field. As early as 1901, the Libby Oil Company brought in a small well at 1,000 feet, but it had no commercial value. Only when the Paraffine Oil Company's Fee #1 roared in on October 31, 1903, did Batson establish itself as a worthy companion of other Gulf coast fields. Fee #1 flowed 600 barrels daily from a depth of 790 feet, with an oil-producing stratum, or pay horizon, of 35 feet. On December 23, Fee #3 presented its owners with the handsome Christmas gift of 15,000 barrels a day.

Because of the uncertainty of whether any well would strike oil, managers prepared storage facilities only in relation to the volume of flow. In places like Batson, huge earthen pits or sumps, around fifteen feet deep and a hundred yards square, served as reservoirs. A foot or two of water covered the bottom and prevented oil from seeping into the earth, since it floated atop the water. At Batson and elsewhere, these pits were often covered with planks to prevent evaporation. The air space below the planks served as a perfect trap for volatile gases, needing only a spark to create an explosion and conflagration. Once at Batson lightning struck a pit, and the fire spread to nine neighboring sumps, with a total loss of 2.75 million barrels of oil.

Like Sour Lake and Batson, Saratoga was a Hardin County salt dome golconda. Big production began in Saratoga in 1902. *Library of Congress*.

Between Saratoga and Sour Lake lay the Texas Company's self-styled Wild Cat Camp, 1909–1910. Perched in the tent entrance is Max T. Schlicher. *UT Archives*.

These 1910 oil-field folk of Saratoga appear more interested in the photographer than in their stump speaker on the right. *Sun Oil Co.*

Before oil was struck at Sour Lake, the well-to-do went there to take the healing waters. They could be quartered comfortably in the Sour Lake Hotel. *Texas Co.*

Walter B. Sharp, 1870–1913, entered the Texas oil industry drilling an experimental well at Spindletop in 1893. In 1902 he earned a fortune for the Texas Company in Sour Lake with his skillful drilling. *Fuel Oil Journal*.

After the boom began in 1902, most residents lived in these more modest quarters. The gentility of the Sour Lake Hotel was eclipsed when roughshod roustabouts took over the town. The Bluefront Hotel (*far right*) was more their style than the elegance of the older hostelry. *Texas Co.*

Those who couldn't afford the Bluefront camped in the outskirts of Sour Lake in 1903. Workers quickly abandoned such makeshift palmetto huts when better quarters became available. *UT Archives.*

Sour Lake mud, almost a universal in drilling operations. *Exxon Corporation*.

Rotary rig in Sour Lake. *American Petroleum Institute*.

Wreck of Guffey #2
July 11-02.
Sour Lake Tex

Photo
by H L E Agerton

The J. M. Guffey Company invested heavily in the Sour Lake field. Guffey #2 proved a loss. *Exxon Corporation.*

Huge equipment, such as this boiler, had to be hauled through the Sour Lake backwoods by oxen. *Texas Co.*

Fee #3 at Sour Lake was completed in January, 1903, as the Texas Company's first producing well. *UT Archives.*

The Sour Lake "Shoestring" district, or Veech tract, the richest part of the field. The dimensions of lots for sale there were 30' × 1,200'. Three rows of wells covered the Veech 1,200 acres. In 1929 a sink swallowed about thirty acres of the Sour Lake field, including part of the Shoestring. *Corsicana Public Library*.

A Sour Lake drilling crew pauses for the camera in 1903. *Texas Mid-Continent Oil & Gas Association*.

Oil fire at Sour Lake that destroyed over 100,000 barrels. The original caption, with perhaps unintentional humor, noted that Texas has "oil to burn." *Texas State Archives*.

Saratoga, Sour Lake, and Batson / 75

A BATSON OIL GUSHER

The second well drilled in Batson was the first brought in by the Guffey Company in late December, 1903. Poisonous gas from the well forced residents of Batson to move hastily in the middle of the night of January 7 and killed five men, plus many horses, hogs, and chickens. The well was then sealed. Early in February the derrick burned, but it was immediately rebuilt. On March 14 the well was agitated and allowed to gush a few moments to clear out mud and sand before oil was turned into tanks. While it was gushing, this picture was taken—the first gusher photographed in Batson. *Corsicana Public Library*.

*Left*: The Pride of Batson! *Southwest Collection*. *Right*: Choate #2 well of the J. M. Guffey Petroleum Company at Batson, 1906. *Left to right*: Harry Wilson, Ezra Frezay, G. W. Holleyman, Ed Summerrow, and Martin Rowe. *UT Archives*.

Batson's Oriole gusher gave cause for celebration on the Fourth of July, 1904. *Southwest Collection*.

Saratoga, Sour Lake, and Batson  /  77

For want of a jailhouse, Batson authorities chained prisoners to trees. This picture was probably posed. *UT Archives*.

A family outing for Guffey's employees. In Batson, as in most boom towns, social virtue battled with vice. *UT Archives*.

The wages of sin. Every Monday morning arrested whores and gamblers appeared before the JP in the courtroom on the Crosby House's second floor. The JP set fines large enough to provide income for him, the sheriff, and the arresting officer. He then displayed the women on the balcony, with expectant men gathered below. Each woman's fine was announced, and the first man to reach the court and pay her fine could keep her for twenty-four hours. *UT Archives.*

A rare shot of two wells gushing simultaneously at Humble. *Exxon Corporation*.

# Humble

LIKE the other salt dome fields, that at Humble, about twenty miles north of Houston, had indicated the presence of oil by gas seeps. For many years prospectors had thought it might pay to drill the area. C. E. Barrett of Houston drilled several shallow test wells in the summer of 1904, with disappointing results. In October the Higgins Oil and Fuel Company brought in a large gas well one-half mile from Barrett's. Then early in 1905 Beatty #2 began to produce 8,500 barrels daily from 700 feet. That well proved the possibilities of the Humble field, and a rash of drilling ensued. In the following three months, Humble yielded 3 million barrels.

In the flush days of the Humble field, a notable pair participated in the venture, Walter B. Sharp and Howard R. Hughes. As a result of his drilling experiences there, Hughes began work on a rolling bit better suited for rotary operations than those currently employed. In 1909 he devised a rolling bit with two cones whose teeth were spaced unevenly. As they rolled with the rotary motion of the drill, these hard steel cones ate into the rock. Because the conical bit could drill a straight, round hole quickly and efficiently, it was vastly superior to the older fishtail bit. The latter would wear rapidly because rocks dulled its cutting edges. When a fishtail bit became dulled, it had to be removed and dressed. On the drilling crews of early wells, tool dressers kept their forges going and the rings of their hammers could be heard through the field as they placed heated bits on the anvil and pounded them back to the proper dimensions. After Hughes invented the cone bit, the Sharp-Hughes Tool Company manufactured it, with considerable profit. Before Sharp died in 1913, Hughes had bought out his interest and the Houston firm was known as the Hughes Tool Company.

In 1917 Hughes improved his bit, increasing its drilling efficiency and lessening costs. The reaming cone bit had a longer head with two hardened steel rollers mounted on perpendicular shafts. The rollers had teeth of different pitch that kept the hole reamed out evenly.

Another occurrence that distinguished the Humble field was that it lent its name to a company that began operations there, the Humble Oil Company. The field took its name from the nearby village of seven hundred that had in turn been named for Mr. Pleasant Humble, a justice of the peace. Larger than most Gulf coast fields, Humble was about two and one-quarter miles wide and four miles long. Many companies drilled there, including the Paraffine Company that luckily had leased the fabulous "Paraffine forty acres." Unlike production in other salt dome fields, that at Humble was from five different pay sands, ranging from 600 to 3,200 feet. In 1917 oil was discovered at much deeper levels, revitalizing the field.

Social conditions in Humble resembled those elsewhere along the Gulf coast. Oil-field toughs and sharpies quickly turned callow farm boys into wary, self-sufficient men. Fighting and killing were frequent, with the dead usually getting summary oil-field burials. Sanitation facilities did not exist, but hordes of mosquitoes did. Workers drew wages of three dollars for a twelve-hour day. Their expenses included fifteen cents for a drink of whisky or a bottle of beer; twenty-five cents per haircut; fifteen cents for a shave; fifty cents each for trousers and shirt; twenty-five cents per bath; a dollar a day for room and board. Saloons, gambling houses, and bawdyhouses eagerly fleeced roughnecks of their hard-earned money. A Humble saloon, constructed of slabs from a nearby sawmill, offended particularly. If workers did not spend freely, the bartender would get them drunk, haul them to the back room, and lift their wallets. Growing tired of the trimming, some of the boys decided to retaliate. A railroad ran alongside the ginmill. One night the boys attached a strong bailing cable to a corner stud of the building. Then when a logging train crept by, they fastened the cable to one of the cars "and scattered that slab saloon about four hundred yards right down the railroad track."

Concurrent with production along the Gulf coast, oilmen built refineries near markets and transportation centers. In some cases the refineries were located near oil fields, but frequently it was more economical for them to be close to cheap water transportation. Texas City, with its ideal port facilities on Galveston Bay, had a refinery as early as 1912.

Moonshine Producer's #10, 1904, the year oil was struck in Humble. *UT Archives*.

Note the roofed section of this otherwise open sump. Compare with the photograph of the open sump at Spindletop on page 47. *Corsicana Public Library*.

TEXAS. UNDERGROUND TANKS ON FIRE HUMBLE.TEX

Open sumps were particularly vulnerable to this disaster. *Corsicana Public Library*.

First double Deck Derrick

Harry R Decker

SHARP & HUGHES HERMAN N°4 HUMBLE TEX, APRIL—1906

RST OIL WELL EVER DRILLED UNDER CONTROLLED PRESSURE
ntract for this Well called for: Success-drilling double pay — Failure no pay

In addition to the firsts inscribed on this picture and the incentive for success, names of those who financed the drilling are important. Sharp's widow financed the Oral History of Texas Oil Pioneers that collected photographs, including this one, in addition to interviewing early oil-field workers. Howard R. Hughes, Sr., invented the cone bit in 1909. The famous Hughes Tool Company was an outgrowth of the Sharp-Hughes Tool Company. H. R. Hughes was the father of Howard, aviation and movie pioneer and later an enormously rich, eccentric recluse. *Hughes Tool Co.*

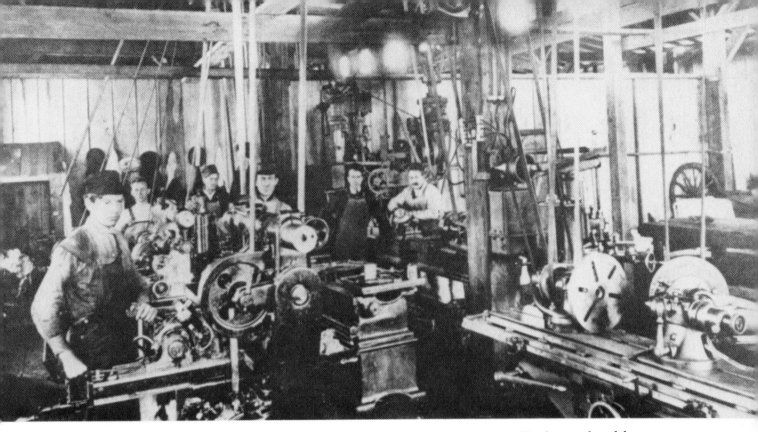

The factory where H. R. Hughes began to make his fortune. The cone roller bit produced here revolutionized rotary drilling. This bit drilled a straight, round hole with great speed and efficiency. In addition, Hughes Tool Company manufactured tool joints, valves, and other oil-field equipment. *UT Archives*.

*Left*: The young Hughes, taken around the time he invented the cone bit. *John H. Lindsey, Houston*.
*Right*: Hughes continued to improve drilling bits. In his Houston laboratory, this tricone is being tested in a rock formation. *Hughes Tool Co.*

Humble / 85

After the picture, the crew ran this Hughes milling cutter rock bit into the hole, 1917. Note that the meshing of the cogs on the drive shaft and rotary table is protected by a safety shield. *Hughes Tool Co.*

In this 1917 photograph, the crew has just pulled the drill string from the hole. No shield protects these workers from the meshing cogs. *Hughes Tool Co.*

In 1917 Hughes experimented with a tunneling machine to drill from Allied trenches through no man's land to German trenches. The white shirt and tie symbolized his being the boss, although his pants obviously got dirty. *Hughes Tool Co.*

In business attire, Mr. Hughes stands outside the tent housing the experimental tunneling machine. *Hughes Tool Co.*

A reamer and roller bit used in 1920. *Southwest Collection.*

A rotary drill in the Humble field. Note the cluster of wells. *Smithsonian Institution*.

In Humble, this well flowed directly into its own sump. *Exxon Corporation*.

Frequently several close wells would share an open storage pit like this one in Humble. On a calm day, such as February 23, 1916, the sump formed a perfect mirror. *Exxon Corporation*.

By 1917 the Humble field was more the province of well-dressed businessmen than of the wool-hatted, gallused roughnecks in the photograph on page 83. *Houston Public Library*.

The Pierce-Fordyce Oil Association refinery at Texas City, 1912. Ocean-going tankers could come to the Texas City docks on Galveston Bay. The company also had a refinery in Fort Worth at this time. *Texas Mid-Continent Oil & Gas Association*.

This 1923 aerial photograph shows a tank farm and docks at Texas City. The grain elevator indicates a diversified economy. *National Archives*.

At work in the Burkburnett field. Such work made this book possible. *Exxon Corporation*.

# Red River Uplift: Petrolia, Electra, Iowa Park, Wichita Falls, and Burkburnett

AFTER the Gulf coast fields, the next important area of discovery was North Texas in counties bordering the Red River. Geologists called the oil-bearing formation the Red River Uplift. As early as 1900 at Petrolia in Clay County observers had suspected that oil lay near the surface. The village had taken its name, in fact, from gas seeps nearby. In 1902 on the Lochridge farm near Petrolia, drillers of a water well discovered at 150 feet a true oil sand that produced a modest amount of oil. Two years later some eighty wells were completed in the area, yielding between three and forty barrels daily, so the field held little excitement in comparison with those of the Gulf coast.

The Petrolia field did become significant, however, with the completion of natural gas wells. The Navarro Refining Company drilled the first of these, tapping gas at 1,500 feet. E. R. Brown, brought by J. S. Cullinan to Corsicana to construct the refinery, headed the Navarro Refining Company, as well as the Clayco Oil and Pipe Line Company and the Corsicana Petroleum Company. When this gas was piped to Wichita Falls, nineteen miles distant, for commercial use, the Texas natural gas industry was born. Brown, with the backing of Henry Clay Folger and Calvin N. Payne, got the Lone Star Gas Company chartered in 1909, with the initial purpose of marketing gas from the Petrolia field. In 1910 it piped gas into Fort Worth and Dallas. A decade later fifty-nine wells were furnishing gas for 133 industrial and 14,719 domestic customers in the state.

A real boom came in the Red River Uplift in 1911 when the Electra field became a major producer. As with many other Texas fields, this one first came to notice after the turn of the century as a result of drilling for water. On his ranch thirty miles west of Petrolia, W. T. Waggoner sought artesian wells to water his cattle. One of these wells was near Beaver Switch, a stock-shipping depot of the Fort Worth and Denver Railroad. That well offered only traces of oil and plenty of salt water. Later, when oil production centered around Beaver Switch, Waggoner changed its name to Electra, after his daughter. (Apparently his wife was no student of Greek mythology!) Oil on the Waggoner property resulted in a great fortune for the owner, but at the time he discovered it, he feigned disgust that oil had spoiled his water well. "I wanted water, and they got me oil. I tell you, I was mad, mad clean through. We needed water to drink ourselves and for our cattle to drink. I said, 'damn the oil, I want water.'"

The Producers Oil Company, a subsidiary of the Texas Company, drilled a number of dry holes at Electra before striking oil. Its Waggoner #5 hit pay at 1,852 feet on January 17, 1911, to become a fifty-barrel-a-day producer. The company sought to withhold news of the strike, setting up an armed guard at the well and erecting a wire fence around the site, measures hardly taken with dry holes! Probably because of these measures, word got around quickly that Electra had a producing well.

Real excitement came when Clayco #1 gushed one hundred feet into the air on April 1, 1911. Two and one-half miles north of Electra, this well ushered in a genuine boom. As soon as the news spread, oil speculators converged on the town from Louisiana, South Texas, and Tulsa. A tent city mushroomed, and Electra experienced all the stimulation and difficulties of a town bitten by the oil bug. In the first nine months, the field's 101 wells yielded 892,204 barrels of crude, all but 70,000 purchased by Magnolia.

In addition to having an excellent oil field, Electra experienced other advantages. Already located on the Fort Worth and Denver Railroad, it had fine transportation. A special shuttle train, "Coal Oil Johnny," plied the twenty-seven miles between Wichita Falls and Electra, hauling workers, speculators, and spectators to the field. The FW&D also offered fine tank-car service. Four pipelines were laid, primarily to the Fort Worth–Dallas area. As the field continued to produce, the town developed as a small oil center, with the usual institutions, a refinery, oil-field supply firms, a tank manufacturer, and machine and welding shops.

Situated as it was between Petrolia and Electra, Wichita Falls wanted desperately to be more than a warehouse for the oil towns. It aspired to be a proper oil town, too. To promote this goal, it offered an incentive of $5,000 for the first well within six miles of the city limits. However keen the desire, Wichita Falls gained prominence not as an oil field but as the regional headquarters for the production of the Red River Uplift. In addition to becoming an equipment supplier for the fields, it had refineries, an indoor and outdoor stock exchange, and other connected businesses.

Adding to the region's growing fame as an oil center, the Iowa Park field, twenty miles west of Wichita Falls, began producing on January 27, 1913. On that date the Forest Oil Company struck oil at 486 feet on the Ferguson farm, with a flow of twenty barrels daily. Other wells came in at shallow depths of less than 1,000 feet but with characteristically small yields. Nonetheless, Iowa Park produced steadily for years.

A dozen miles north of Wichita Falls, hard by the Red River, Burkburnett faced its destiny as the prototypical boom town. At least, Hollywood bestowed this destiny upon it in 1940 by filming *Boom Town* in Burkburnett. The movie starred Hedy Lamarr, Claudette Colbert, Spencer Tracy, and Clark Gable. The initial stirrings, however, did not presage such fame. When the Corsicana Petroleum Company's Schmoker #1 hit oil at 1,837 feet on July 8, 1912, no wild jubilation occurred over its eighty-five barrels a day. Chris Schmoker, the old farmer on whose land the well was located, was happy enough to have a little additional income, but this was not a new Spindletop. For five years, the Burkburnett field counted as just another middling producer, economically worthwhile but hardly exciting.

Then came "Fowler's Folly." S. L. Fowler owned a farm just north of Burkburnett and had about decided to sell out and move on in typical frontiering tradition. His wife had been infected by the oil bug so active within a twenty-mile radius and induced him to stay and try to find oil. After studying the neighboring land where wells produced and comparing his own, he agreed to follow his wife's advice. With three partners, Movel Fowler, W. D. Cline, and J. A. Staley, he began the quest. Fowler furnished his land and $500, and his brother Movel invested another $500. Cline, a wildcatter, contributed his cable tool rig, his services, plus those of his tool pusher in exchange for $1,000 stock in the venture. Staley invested $500. Their agreement specified that they would drill up to 1,700 feet and that if they found commercially significant oil or gas, they would form the Fowler Oil Company, capitalized at $12,000.

On July 29, 1918, "Fowler's Folly" erupted with a daily flow of 2,200 barrels. Its pay sand lay at 1,734 feet. Luckily Cline had been willing to gamble on a few extra feet. And what a handsome gamble it was! After the company drilled a second producer on Fowler's farm, it sold out to the Magnolia Petroleum Company for $1.8 million. President E. R. Brown of Magnolia paid the Fowler Company by check, the first for $800,000 and two subsequent ones of $500,000 each.

Within three weeks of the discovery of Fowler's well, fifty-six derricks jutted into the Burkburnett sky, attesting to the faith oilmen now had in the field. Both Magnolia and the Texas Company ran pipelines to the producing wells, eager to collect the crude. The Burkburnett field quickly became a maze of derricks, storage tanks, above-ground piplines, boilers, and miscellaneous equipment, such as bull wheels. Interspersed throughout the field were living quarters—tents, tarpaper shacks, shanties, and bungalows that obviously predated the discovery of oil. Photographs show how completely intermingled were home and work in Burkburnett. They also demonstrate that homeowners made the most of their opportunities. Not only were they eager for wells to be drilled on their town lots, but they also erected tents or built lean-tos as rental property. They realized that booms could be short-lived and gladly made the most of their opportunities.

As in all boom towns, swarms of people milled about Burkburnett. The train station always attracted folks. Because of a housing shortage in town, many oil-field workers lived in Wichita Falls and rode the morning and afternoon commuter train to and from work. Others gathered at the depot to receive drilling equipment and various consumer merchandise popular in the boom town. Lease hawkers filled the streets and stores, buying and selling continuously. Before the boom, one citizen had tried to sell his house and lot for $1,500, but without success. Afterward, he leased a portion of the lot for $3,600, retaining part of the royalties. It was not uncommon that leases on small lots would be quickly traded, sometimes with a tenfold profit. Traders had to be wary, since all boom towns attracted dishonest speculators. Many a greenhorn found himself holding worthless paper. A Wichita Falls newspaper reported on August 2 that throngs of visitors walked "the streets all day and until late at night. . . . To say that Burkburnett presents a busy scene does not begin to express conditions in that little city, where all is activity and where the main street is thronged with visitors from both near-by and distant points. . . . Oil companies are being formed on the sidewalk or in the middle of the street." Since prohibition already prevailed in Texas, a flourishing bootleg business sprang up in "blind tiger" bars. These bars operated so that the customers and dispensers of spirits could not see each other. The bartender stood behind an opening like a bank teller's window, except without the window. The only aperture was a slot from six to nine inches high. The customer laid his money there and a disembodied hand provided the bootleg liquor. Such recreation frequently operated in connection with another popular, but illegal, diversion—prostitution. Naturally, boom towns offered ready-made targets for the Texas Rangers. Even in a boom town like Burkburnett, however, the spirit of gallantry never died completely. Early in 1920 after two heavy snowfalls, streets turned into quagmires. Men in high rubber boots offered to carry women across such street for ten cents a ride.

The Burkburnett field was well established when further exploration revealed vast quantities of oil in its northwest extension. On April 29, 1919, the Burk-Waggoner Company struck oil at 1,705 feet, the well producing 2,750 barrels daily. This well set off a new spree of speculation and stock manipulation. Drillers followed the rule of capture, putting down wells as close together as possible, even into the Red River. In 1919 the Burkburnett field, including the northwest extension, called Newtown, yielded an astounding 31,604,183 barrels of crude. From the time Schmoker #1 began flowing in 1912 through 1918, Burkburnett had produced approximately 8,400,000 barrels, so 1919 constituted a genuine banner year.

As of January, 1919, 289 incorporated companies operated in Burkburnett, most recently organized. The field boasted 168 flowing wells, with 207 under way. Unproved land sold for $180,000 per acre. Twelve pipeline corporations handled oil from the field, including such majors as Gulf, Magnolia, Humble, and the Texas Company. Since Burkburnett had three refineries, not all the oil had to leave the immediate area for manufacturing. Those three, plus nine at Wichita Falls, refined 27,800 barrels of crude oil a day. Burkburnett and other fields in the Red River Uplift likewise furnished crude for seven refineries in Fort Worth, two in Iowa Park, and four in Dallas.

1904 brought the oil boom to North Texas with this first gusher at Petrolia in Clay County. Petrolia is nineteen miles northwest of Wichita Falls and six miles from the Red River. *Wichita Falls Times.*

A steam tractor and mule team combine to haul a 33-ton main engine frame to the Lone Star Gas Company station in Petrolia, 1914. *American Petroleum Institute*.

Drilling crew for Clayco #1, two and one-half miles north of Electra. *Left to right*: Hal Hughes, unidentified, Clabe Moody, and S. C. ("Dad") Massengill. When the well came in on April 1, 1911, spouting oil 100 feet high, the Electra boom began. By the end of 1911, the Electra field had produced 1,402,802 barrels. *Southwest Collection*.

A Star drilling machine at Electra. *Southwest Collection*.

A Star spudder used to begin a well. *Permian Basin Petroleum Museum*.

A gusher breaks the monotony of the Electra landscape. Note the network of pipes in the foreground and the stack of pipes to the left of the well. *Exxon Corporation*.

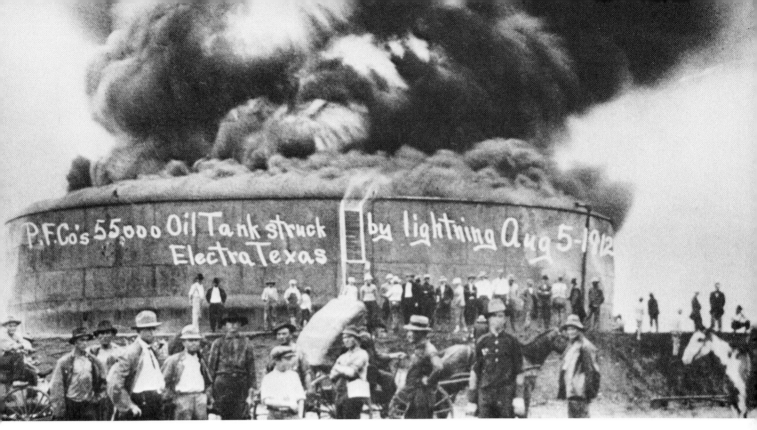

Natural disaster struck this Pierce-Fordyce tank, attracting those more curious than prudent. On May 2, 1914, the Texas Company's #240, another 55,000-barrel tank in Electra, burned up. *UT Archives*.

Oil-field folk of Electra paying last respects to the wife of a carpenter who erected derricks. *Southwest Collection*.

From the derrick floor, this gusher at Wichita Falls in 1919 looks as good to the white-collar managerial types as to the drilling crew. *Fort Worth Star-Telegram*.

Electra headquarters for one of the early giants of the industry, 1911–1912. *Seated, left to right*: W. A. Sniffin and O. C. Baker. *Standing*: S. R. O'Day, B. M. Dickey, George Lewis, unidentified, Frank M. Baker, E. N. Wilson, and two unidentified. *Southwest Collection*.

The Wichita Falls Central Stock Exchange, 1919. These traders in securities leases didn't work as hard as the roustabouts to earn (or lose) money in the oil business. Some appeared to be acquainted with field work, such as the seventh from the right in the front row. *Wichita Falls Times*.

Trading in oil leases and shares was as lively on the curb in Wichita Falls as indoors. Company names, in addition to Golden Rule, included Over the Top, Block Five, Sam's Clover Leaf, Bit Hit, and O Boy! *American Petroleum Institute*.

Wichita Falls' Panhandle Boiler Factory and refinery, 1930. *Library of Congress*.

An Archer County well blows over the top on March 29, 1931, demonstrating the continuing productivity of the Red River Uplift. *Fort Worth Star-Telegram*.

Burkburnett in April, 1919, at the height of the boom. Wells are identified on the photograph as (*left to right*): Russell Sanderson; Russell Sanderson; Humble Oil Co.; Humble Oil Co.; Weowna Oil Co.; Plainsview Littlefield #1; Perry Browning Oil Co.; Parks Oil Co.; Imperial Petroleum Co. (*foreground*); Perry Browning Oil Co.; Perry Browning Oil Co.; Paris Oil Co.; Security Oil Co.; Cullinan Oil Assn. (*foreground*); Mayflower Oil Co.; Burk Burnett Home Oil Co.; Burk Burnett Home Oil Co.; Cooper & Meyers; Ozark Trail #2; Cass Oil Co.; Block 34; HooHoo Oil Co.; Bank Oil Co.; Bank Oil Co.; Heinie Meyers; Big 3 #1; Big 3 #4; Double Burk; Weowna Oil Co.; Lone Star Oil Co. #1; Lone Star Oil Co. #2; Drillers Oil & Gas #2; Morris Oil Co.; Drillers Oil & Gas; National Oil & Gas Co.; Sunflower Oil Co.; Narona Burk #1; Byrens Oil Co. #1; Pickle #1; Narona Burk #2; Crowell & Thompson #1; McGee #2; Pullman Oil Co.; McGee Oil Co. #2; Abner Davis #5; Over the Top; Thorn Oil Co.; Vernon Oil Co.; Big Hit Oil Co.; Robertson Petroleum Co., Pickle #2; Rogers & Thomas; Frederick #1; Crowell & Thompson #2; Frederick #2; Panhandle Oil Co.; Centerfield Oil Co.; Hale & Todd; Parker & Anderson; Texahoma Production Co.; Hale & Todd; School Block #1; Hale & Todd #3; Hale & Todd #1; Chams Oil Co. #1; Chams Oil Co. #2; Abner Davis #4; Ganado Oil Co.; Lula Park Oil Co.; School Block #4; Herman Morris Oil Co.; Imperial Petroleum Co.; Sunbeam Oil Co.; Waggoner #1; Jane Louise Oil Co.; Great Dome #1; Burk Burnett O'Neil #1; McElroy Oil Assn.; Rowe, Stayley & Blair; Brown #2; Lesh Oil & Gas Co.; Middle Buster; Edna Burk; Apex #2; Great Dome #2. *Humanities Research Center, UT*.

A drilling machine and tent colony among the mesquite trees on the Watkins lease outside Iowa Park (twenty miles west of Wichita Falls), April 18, 1919. *Humanities Research Center, UT*.

Crew for the first Burkburnett well, known as Corsicana Petroleum #1 or Schmoker #1. *At left*, Elmore Trigleth. *Standing*, an officer of the Corsicana Petroleum Company (later Magnolia) that drilled the well. *Right*, Harry McKinley. Others unknown. Accounts vary as to the date the well began producing: July 1 (as above), 4, or 8, 1912. Photographers' inscriptions may mislead. *Wichita Falls Times*.

*Left*: Chris Schmoker, owner of the farm where the Burkburnett boom began in 1912. *Fort Worth Star-Telegram*. *Right*: Beyond Burkburnett's Texas Chief flows the Red River. *Library of Congress*.

In the northwest extension of the Burkburnett field, Burk-Waggoner #1 started flowing on April 29, 1919. Gas pressure was sufficient to spew oil into the pit. *Library of Congress*.

Red River Uplift  /  109

*Left*: The horse seems unexcited by another gusher in the Burkburnett northwest extension. *American Petroleum Institute*. *Right*: Observers look as if they are calculating royalties as oil spews into a sump in the Burkburnett field. *Library of Congress*.

Space was at a premium on the train between Wichita Falls and Burkburnett in 1919, as men sought their fortune in the oil field. Those who couldn't get aboard could have walked the dozen miles. *Wichita Falls Times*.

Burkburnett, Boom Town, U.S.A. *Wichita Falls Times*.

The Wichita Falls photographer didn't minimize the import of the Burkburnett field in 1919. With his store on wheels, the merchant in the foreground could set up shop wherever business seemed best. *American Petroleum Institute*.

Some houses in the Burkburnett field predated the boom, but shacks and tents accommodated the influx of workers. Money was made by landlords as well as lease owners. *American Petroleum Institute*.

Residential Burkburnett. *American Petroleum Institute*.

Burkburnett, 1918. Note the housing around the boiler in the foreground. *Southwest Collection*.

In the petroleum industry, stills were used to refine crude oil. This sheriff's posse in Burkburnett rounded up stills used differently by oil-field workers. *Wichita Falls Times*.

Ladies' Day in the Newtown (Burkburnett) field, 1919. *Bottom row, third from left*: Luke Erwin; *fifth from left*: H. A. Erwin. P. L. Gilbert, driller, is seated on pipe flowing oil. *Second row, third from left*: A. N. ("Andy") West. *Wichita Falls Times.*

Burkburnett's main street was ready to receive the influx of workers. Note the business in the left foreground with the sign, "Baggage Checked Here." *Fort Worth Star-Telegram.*

Drowned out on Burkburnett's main street, 1919. *American Petroleum Institute.*

In Newtown, the oil-field section of Burkburnett, a little rain made the thoroughfares impassable. *Exxon Corporation.*

Even if Burkburnett farmers struck oil, they didn't forget their first calling, as this harvested field shows. *Exxon Corporation.*

Tank cars at loading platforms in Burkburnett. They transported crude oil to refineries in Wichita Falls, Fort Worth, and elsewhere. *Southwest Collection.*

Offshore drilling in the Red River, 1920. The Bridgetown field was an extension of the Burkburnett field, located where the Grandfield Bridge crossed the river into Oklahoma. *Southwest Collection*.

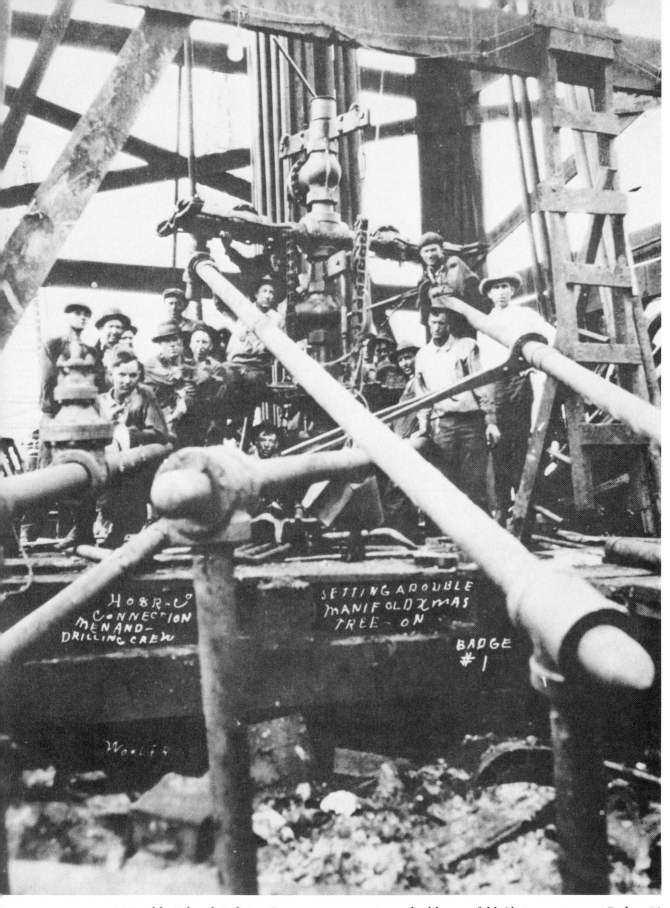

A Humble Oil and Refining Company crew setting a double manifold Christmas tree on Badge #1. *Exxon Corporation.*

# Goose Creek–Baytown

A distinctive feature of the Goose Creek field was that it encompassed both land and water. The first offshore drilling in the United States was in the Santa Barbara, California, channel in 1898. (Wells in that channel once again jumped into the headlines in 1969 when they leaked great quantities of oil into the water, killing marine life and sea birds and spoiling superb beaches. That incident seemed to mark a turning point in the public's tolerance of environmental pollution by oil companies. Clearly, the petroleum industry has been on the defensive on this issue since then.) Goose Creek claims a first for offshore drilling along the Texas coast. The initial strike there on June 2, 1908, was so close to the bay that further exploration under water seemed logical. That discovery well, drilled by a Houston syndicate, hit oil at 1,600 feet, intermittently producing about thirty barrels per hour. Such a yield was insufficient to create further interest in the field.

At least one owner of land near the producing well refused to believe that no more oil lay beneath the surface. In the summer of 1916, John Gaillard fished regularly in Tabbs Bay. Noticing bubbles floating to the surface, he surmised that a school of buffalo were feeding on the bottom. When he kept seeing bubbles from the same spot on successive days, he concluded that he was fishing over a gas seep, not a school of fish. He contracted with Charles Mitchell to drill a test well on his property close to the shore, 500 feet west of the 1908 well. On August 23, 1916, Mitchell brought in a gusher from 2,017 feet. The 8,000 barrels daily heralded an important new field, and the prosperity of Goose Creek began with that discovery.

The Gaillard well slumped to 300 barrels a day within two months, but a number of other wells on and offshore lifted the field's production from 700 barrels daily in August to 5,000 by the end of the year. Contributing to this figure was C. T. Rucker's well that produced 6,000 barrels initially from a 2,000-foot sand. During 1917 drilling continued, with 62 percent of the producing wells yielding an average initial flow of 1,181 barrels a day. In March the field's output was 31,350 barrels and in September, 66,000, with steady production during the months in between. The big bonanza of

1917 was the Sweet #11 of the Simms-Sinclair Company that blew in on August 4, with 35,000 barrels daily from 3,050 feet. This well gushed for three days, sanded up, and, when finally cleaned out in March, 1918, yielded 1,500 barrels a day.

With the United States fully involved in World War I during 1918, the demand for petroleum products soared. The United States supplied most of the Allies' needs for oil, and as Lord Curzon stated, "The Allies floated to victory on a sea of American oil." The pressing need, plus high wartime prices of $1.35 per barrel, induced companies operating in Goose Creek to test John Gaillard's theory that oil could be found beneath bay waters. The Humble Oil and Refining Company and Gulf Production Company essayed offshore wells. When these initial efforts proved successful, scores of other wells went down into the shallow water of Tabbs Bay. The field's output for 1918 was 8,943,635 barrels.

Fairly early in the productive history of the Goose Creek field, it attracted the attention of Ross S. Sterling, a founder of the Humble Company and governor of the state from 1931 to 1933. Born in Anahuac, about twenty miles across Trinity Bay from Goose Creek, he became a merchant in Galveston and later in Sour Lake. There he sold feed for draft animals used in the oil field. That succeeded so well that he opened similar stores in Saratoga, Humble, and Dayton. Building upon that capital accumulation, he opened banks in towns where he had feed stores. Then in 1909 he decided, like Mark Hopkins and H. A. W. Tabor before him, to shift from being storekeeper for prospectors to becoming a prospector himself. (Hopkins grew rich from the 1849 gold rush and became one of California's "Big Four." Tabor was the "silver king" of Colorado, its senator, and husband of Baby Doe.) Sterling first purchased oil properties in the Humble field. Gaining experience there, he concluded that small operators suffered from excessive competition—too many wells being drilled. The smart thing would be to form one company that could husband capital and effort. Led by Sterling, G. Clint Wood, who guided Sterling's first purchase of two flowing wells in the Humble field; J. W. Fincher, cashier of Sterling's Humble State Bank; M. C. Hale, stockholder in the Humble bank and a rig-builder; S. K. Warrener, a former gauger for the Sun Oil Company; and Charles B. Goddard, a successful drilling contractor and operator and Warrener's brother-in-law, incorporated the Humble Oil Company in January, 1911. Later that year the U.S. Supreme Court declared the Standard Oil Company a monopoly and ordered its division. Capitalized at $150,000, the Humble company listed producing properties as its assets. The formula for determining the worth of a well was $350 per barrel daily. Since the founders operated in the Humble field, they named their company after the field. In 1912 it moved its headquarters to Houston.

From the time Ross Sterling bought the Southern Pipe Line Company for Humble in January, 1917, the histories of Goose Creek, as a producing area, and the Humble company became intertwined. Sterling purchased the pipeline to gather crude from the Goose Creek field and route it to the Houston ship channel. Two 7,000-foot lines of four-inch pipe crossed the bay to Hog Island. There they drained into cypress storage tanks holding 20,000 barrels. Ocean-going tankers would dock at the wharf on Hog Island and receive their crude from a 500-foot, six-inch line from the cypress tanks. For a while the Southern Pipe Line Company handled most of Goose Creek's production, but major companies also piped oil from the field, primarily to their refineries in the Beaumont–Port Arthur area.

Shortly after Sterling brought Humble into Goose Creek via the pipeline company, he set in motion forces that would put Humble in the area permanently. On May 17, 1917, he and some associates requested from the Texas secretary of state a charter for the Humble Oil and Refining Company, a request granted on June 21, 1917. The first board of directors consisted of nine men, all important in the development of Texas' early oil industry. Ross Sterling became president, and his brother F. P. was a vice-president. Other vice-presidents included William S. Farish and Robert L. Blaffer, Spindletop veterans who helped develop fields in the Red River Uplift; Harry C. Wiess, whose father had rented mules to Captain Lucas so he could dig a sump for the Spindletop gusher; and Walter W. Fondren, a driller in the Corsicana field, later distinguished for philanthropy to Southern Methodist University and Rice University, among other institutions. The directors were Charles B. Goddard, a driller from Spindletop; Lobel A. Carlton, a Beaumont lawyer specializing in petroleum cases; and Jesse H. Jones of Houston, a real estate tycoon, lumberman, and general promoter of Houston's economic welfare. During the New Deal, Jones became director of the Reconstruction Finance Corporation and served as President Roosevelt's secretary of commerce. Farish wanted Jones on the board of directors so that he could "accomplish some financing plans." That Jones was a financier rather than an oilman was attested to by his selling his Humble stock at a considerable profit within a year of the company's organization and resigning from the board.

Through its pipeline operations in Goose Creek, the new Humble Oil and Refining Company became well acquainted with the field and thought highly of its future prospects. So highly, in fact, that it decided to build its major refinery adjacent to the field. Since the field was only thirty miles east of corporate headquarters in Houston, the location of the refinery at Goose Creek seemed preordained. And yet, the Humble bosses stumbled over the place name. According to the Larson and Porter official history of Humble, the refinery could not be put at a town with "so meaningless a name as Goose Creek," so they decided to call the location of their plant Baytown, after the network of bays forming the periphery of Galveston Bay. Naturally, a community grew up around the refinery and the town became known as Baytown.

With the company having decided that Goose Creek was a "meaningless" name, a curious thing occurred. In the early 1920's, three separate municipalities developed: Goose Creek, Pelly, and Baytown. The old Goose Creek field, taking its name from a stream feeding into Black Duck Bay, became part of Pelly. The town of Goose Creek lay to the east of the field and had no oil production within its borders.

Baytown was the home of the refinery, with its docks on the Houston ship channel. Collectively, the area was known as the Tri-Cities until unification in 1946.

At that time citizens debated hotly over whether the new municipality should be known as Goose Creek or Baytown—nobody championing Pelly, which was being called humorously "the tail of two cities." Still the oil-field town, Pelly was not as socially or economically desirable a place to live as the other two. Since Goose Creek was the original community with its celebrated oil field, residents of that town championed its name. Baytowners, who always claimed that Goose Creek really sounded a little tacky and hickish, contended that Baytown would carry more respect. Besides, the refinery was there, and seamen from the world over had come ashore on the Humble docks, en route to Baytown's notorious Main Street brothels. The argument that the world's sailors should not be confused with a name like Goose Creek seemed to carry the day and residents voted in 1948 to live in Baytown thenceforward.

For oldtimers, Goose Creek died hard. True, the school system still bore its name, but more was needed. When the refinery displaced the municipal golf course from its grounds (after 1958 when Humble became a totally owned subsidiary of Standard Oil of New Jersey), a new course was developed—at the Goose Creek Country Club. For many years after unification, students at Baytown's Robert E. Lee High School continued to sing lustily at athletic contests: "Oh, when those Goose Creek Ganders fall in line . . ." (The newer high school in the district is fittingly named for Ross Sterling.)

Prior to the decision to build a refinery at Baytown, Humble had constructed several smaller plants to serve local needs around the state. This one, however, was designed as a complete, modern plant. As early as 1919 Standard Oil of New Jersey had bought controlling interest in Humble stock, so it offered its assistance in constructing the refinery. Despite being the majority stockholder, Jersey Standard chose not to impose its will on Humble. The eastern company recognized the ambition and ability of the Texas concern's directors and their determination to make a respectable profit. As long as the Humble crowd fulfilled that goal, Jersey Standard did not interfere with local management. It treated Humble as a complementary unit in its operations, keeping Humble fully informed of its plans and activities so as to avoid wasteful duplication.

Humble readily accepted Jersey Standard's proffered assistance in erecting the Baytown refinery on a 2,200-acre tract adjacent to the ship channel. Work began on April 16, 1919, when engineers arrived on the marshy ground and started examining their blueprints. Land had to be cleared of pine trees and swamps drained. The latter came as quite a challenge, since deluges kept filling up drained areas. In addition to heavy spring rains, Baytown, as a coastal location, featured omnipresent mosquitoes, lots of grasshoppers, and an assortment of poisonous snakes. Earlier, Brahma cattle had been imported into the area because their thick hides could withstand mosquito bites. These bellicose cattle continued to roam through the refinery grounds, often disrupting the work of terrified laborers.

Late in July the Turner Construction Company was ready to pour concrete. In August work crews began building storage tanks, digging sewers and pipelines within the plant, and laying brick. One difficulty was in securing experienced labor. The Ranger and Desdemona booms had drained off skilled boilermakers and riggers, and many ordinary white construction workers intensely disliked the climatic conditions of Baytown, with its muggy summer days. Since blacks and Mexicans were barred by Texas folkways from oil-field labor, they furnished a work force that Humble could use. Half the laborers constructing the Humble refinery were black or Mexican by the end of 1919, with the Latins predominating. Whatever the difficulties, work proceeded apace, and by the beginning of 1920, machine shops operated, the concrete installations for the filter house, boilerhouse, and crude stills were almost finished, and atmospheric stills were going up. On May 11, 1920, the first oil was pumped into a still. For many years Baytown celebrated that day as Humble Day. The refinery was officially pronounced completed on April 21, 1921, commemorating San Jacinto Day, when in 1836 Texas won its independence from Mexico at the battleground no more than a half-dozen miles from the refinery.

The completed refinery did not operate according to Humble's expectations. One problem concerned the assistance of Jersey Standard personnel, who filled most of the foremen's posts. They were unaccustomed to the locale and did not understand the ways of the inexperienced white workers, recruited mostly from the farm, or those of the Negroes or Mexicans. Then, before operations could become routine, the management decided to add more units to manufacture gasoline, the demand for which was escalating. Added would be sixteen Burton stills, developed by Dr. William Burton in 1912 for Standard Oil of Indiana. The Burton still cracked hydrocarbon molecules into lighter fractions, thus recovering more gasoline from the crude. As higher octanes were needed, the cracking process was advanced with thermal cracking units that put greater heat and pressure on the crude. When completed, the refinery cost over $10 million, the 1918 estimate having been approximately ten times less. The earlier estimate, however, was for a 4,000-barrel daily capacity, whereas the completed plant could handle 10,000.

Earlier suggestions of racial trouble among workers materialized in 1921. The Mexicans and Negroes employed during the construction of the plant were retrained as refinery workers once operations began. When drillers and roustabouts got laid off because of depressed conditions in other oil fields, many of them converged on Baytown to force Humble to fire the blacks and Mexicans and employ them. The superintendent dissuaded white protesters of their notions during a confrontation at the plant gate. Blacks and Mexicans kept their jobs but drew less money than their white counterparts, a tradition of the area and industry.

Humble had placed its refinery in Baytown so that it would have ready access to Goose Creek crude. Because of Jersey Standard's needs, however, different crudes had to be used. The parent company needed a high-grade gasoline and cylinder oil that could best be refined from the paraffin-base North Texas crude, so oil was piped in from Burkburnett and Ranger. The experienced hands from Jersey Standard became stumped in trying to refine the coastal crude, and the refinery was embarrassed in not fulfilling orders. Since the Eastern crew was baffled by the local crude, Humble suffered in competition with Gulf and the Texas Company, whose refineries in Port Arthur turned out saleable oil directly from local crude. Humble's inability in this case caused the imported manager, Clifford M. Husted, to resign, tacitly admitting his failure. When Humble lured Jesse James Harrington away from the Texas Company's Port Arthur refinery and gave him the job of head treater, he solved the problem. In atmospheric stills, he treated lubrication distillates with caustic, "producing beautifully finished lubricating oils instead of mere slop."

During the 1920's the Baytown refinery took several steps that enabled it to become an industrial leader. Under prompting from Jersey Standard, it initiated a research program, employing highly trained chemists and engineers. Through the decade it built new units that enabled it to refine more oil and that more efficiently. These included crude stills and steam stills as well as tube-and-tank cracking coils. The advantage of these cracking coils over Burton stills, which had never efficiently refined Gulf coast crude with its asphalt base and troublesome gas oil, was that they operated continuously rather than in batches. By the beginning of 1926, the plant could refine 50,000 barrels a day, and the operation was as modern and efficient as any in the nation. The previous year it earned its first respectable profit. At the end of the decade the refinery had doubled its capacity to 100,000 barrels daily and had "set the pace for all the Jersey organization in technical improvements."

As it opened its refinery, Humble took specific measures to ensure good labor relations, or at least to let employees know where they stood with the company. Not unexpectedly, Humble took its cue from Jersey Standard. In December, 1920, representatives of management and labor agreed to regular meetings where wages, hours, working conditions, and problems would be discussed. These meetings were called the Joint Conference, and its worker membership was elected annually from the refinery's component divisions. Management appointed an equal number. The Joint Conference adopted an agreement that listed reasons why a worker might be fired as well as his right of appeal. A worker could be dismissed without warning for any one of sixteen reasons or for repeating other violations after warning. With management having equal representation with labor, it easily dominated proceedings, especially since the refinery superintendent presided at meetings.

Management clearly understood that satisfied employees work better than disgruntled ones and that better work meant better profits. In an effort to provide for workers' welfare, the company initiated a training program for employees wishing to advance. At no cost to the workers, management offered chemistry lectures to laboratory workers and courses on foremanship to those interested. In the refinery, the foreman, or pusher, formed the immediate link between workers and management, so a foreman who knew how to handle men was a distinct asset to the company.

Another area in which Humble sought to promote employee welfare was safety. Astounded by its 1920 rate of 2,266 accidents per 1,000 workers, it initiated health and safety committees. The refinery insisted that workers wear metal helmets, safety shoes and goggles, gloves, and gas masks when handling noxious crudes. It also provided first aid training.

Following the tradition of oil companies, Humble provided housing and community facilities for its workers. This tradition began in oil fields distant from towns or where towns did not furnish sufficient housing. If a field was to be developed, a labor force was required, and workers had to have some place to live, with company camps or towns resulting. Humble had previously built camps in the Goose Creek, West Columbia, and Ranger fields. The development of Baytown as a company town proceeded haphazardly. When construction got under way in 1919, Humble's first move was to provide World War I army barracks and tents near the refinery. The company could sleep and feed one thousand workers in barracks and mess halls by January, 1920. It furnished small houses only for married supervisors and skilled hands. Then in 1922 it constructed dozens of one- and two-room rental dwellings for the unskilled. As these houses sat on a lot, their longest dimension was from front to back, and they became known as shotgun houses. You could fire a shotgun at the front door and the pellets would come out the back. The company built schools for white and Mexican children, the latter, DeZavala Elementary School, hard by the plant. For those who wished to see the bright lights of Goose Creek at night, the company offered free bus rides.

In 1923, Humble altered its policy by assisting employees to build homes in Baytown rather than building homes itself. The company laid out tracts, built sidewalks and streets, provided utilities, and offered home loans. When the policy became effective in 1925, a worker made a 10 percent down payment and received from the company a 6 percent loan. This community developed by Humble housed only whites. They also had the use of a fine brick community house, where Humble provided dances and movies in 1925. For decades this community house constituted a focal point for Baytown's social life. It was used for lectures, amateur theatricals, dances, Alcoholics Anonymous meetings, recitals, wedding receptions, and ping pong. Humble offered this facility free of charge to the community, thus ensuring a warm spot in citizens' hearts for "Uncle John," as the company was called.

The stream called Goose Creek gave its name to one of Texas' greatest fields, thirty miles east of Houston. Oil was struck there in 1908, but significant production began eight years later. Since Goose Creek flowed directly into the Houston ship channel, transportation of crude and refined oil was economical. *Houston Public Library.*

Part of the Goose Creek pool lay under Tabbs Bay. On the distant shore is Morgan's Point, where Santa Anna reputedly kept a tryst before the battle of San Jacinto. Morgan's Point is about six miles from the battleground, located at the confluence of San Jacinto River and Buffalo Bayou (which when dredged in 1915 became the Houston ship channel). *Houston Public Library.*

Christmas tree on a Goose Creek well. *Southwest Collection.*

Offshore drilling distinguished Goose Creek from other Texas oil fields, 1919. *Library of Congress*.

Core barrels were used to take samples in a well. From the sample, a driller could tell whether he was close to "pay." *Southwest Collection*.

Wooden storage tanks in the Goose Creek field. *Exxon Corporation.*

Tilt! From these storage tanks at Goose Creek, oil could easily pour onto troubled (or still) waters. *Houston Public Library*.

Offshore production in the shallow waters of the Goose Creek field, 1924. *American Petroleum Institute*.

*Left:* A wooden derrick burns at Goose Creek, June 17, 1924. *Southwest Collection. Right:* Flames toppling a steel derrick. *American Petroleum Institute.*

*Left:* Ross S. Sterling, first president of the Humble Company, at home in Houston, the corporate headquarters. *UT Archives. Right:* Jesse H. Jones, an original member of Humble's board of directors, maintained involvement in the economic and political affairs of Houston, the state, and the nation. *UT Archives.*

Goose Creek—Baytown / 131

On- and offshore wells in the Goose Creek field, 1924. *American Petroleum Institute.*

Baytown's Humble refinery under construction in 1920. The unit shown is filter house #3. Baytown, contiguous with Goose Creek and situated on the Houston ship channel, was an ideal spot for a refinery. Crude oil was immediately available from the Goose Creek field, and refined products could be shipped cheaply by sea. *Southwest Collection.*

The Humble refinery in operation, 1922. *Exxon Corporation.*

Humble gasoline destined for a distant port aboard the tanker *Baytown. Exxon Corporation.*

Tankers at the Shell Deer Park docks on the Houston ship channel. *Shell Oil Co.*

# Other Coastal Fields

SHORTLY after the Goose Creek field became a major producer, other coastal fields within a sixty-mile radius were discovered. The first well in West Columbia, southwest of Goose Creek, dated from 1917, but serious production there came with the Tyndall-Wyoming Oil Company's Hogg #2 on January 4, 1918. The field yielded 186,350 barrels in 1918 and jumped to 5,611,000 the next year. West Columbia's best well was the Texas Company's Abrams #1, completed July 20, 1920. In its first seventy-six days, its oil was valued at $3,825,000. Giant pumping stations piped most of West Columbia's oil to coastal refineries, such as the Texas Company's in Port Arthur.

Only a dozen miles north of Goose Creek lay Barbers Hill, whose most notable family—as the name indicates—were Barbers. In 1889 E. W. Barber dug a shallow water well and noticed gas seepage. After the discovery at Spindletop, the Barbers tried to find oil around the knoll on their property, a formation probably caused by a salt dome. Until 1918, their efforts were fruitless. In that year the United Petroleum Company drilled Fisher #1, striking pay on September 14, 1918. The well yielded only forty barrels daily from 2,142 feet. Further exploration between 1926 and 1930 proved the worth of the field. Ilfrey #2, which began production on February 26, 1930, established an important geological point. Drillers had gone through 878 feet of salt, the toadstool-like overhang of the salt dome, proving that oil existed directly beneath salt formations.

Despite the fact that the Barbers owned much of the land on which oil was found, they were not overwhelmed by their wealth. Like a common-sense agrarian, one of the Barber women, when asked what she planned to do with her money, replied that she just wanted two things: linoleum on the kitchen floor and a new handle in her axe.

Hull, in Liberty County, was the site of a major discovery on July 22, 1918, when the Republic Production Company and Houston Oil Company of Texas brought in their Fee #3 well at 2,352 feet. The well yielded 1,000 barrels daily. Other coastal fields discovered around this time included those at Pierce Junction, Damon Mound, Dayton, Orange, High Island, Markham, Big Creek, and Stratton Ridge.

Because of its excellent dock facilities, Corpus Christi became the natural hub for nearby petroleum production. The area known as the "coastal bend" proved rich in natural gas and oil. As early as 1916, James M. Guffey was drilling in the area and in that year discovered gas near Sinton. Four years later Refugio, twenty-three miles to the northeast of Sinton, developed gas wells. In 1928 oil was discovered there, and by 1930 the Refugio field led the Gulf coast in production. Humble built one of its two refineries at Ingleside, across the bay from Corpus Christi, and in the 1930's other companies built manufacturing plants adjacent to Corpus Christi Bay. Oil and cotton were the major commodities loaded onto ships at the Corpus Christi harbor.

Although Houston had no importance as an oil producing or refining city, it nonetheless played a central role in the coastal oil business. Excepting the Beaumont–Port Arthur area and the Corpus Christi region, Houston became the nexus for petroleum along the broad reaches of the Texas coastal plain. It served as the corporate or regional headquarters for many companies and became the major supplier of oil-field equipment in the area. The ship channel furnished excellent locations for refineries, close to abundant crude supplies and a major population center. Houston did not develop this strategic position in the oil business accidentally—shrewd planners and promoters like Joseph S. Cullinan, Patrick Calhoun, John Henry Kirby, and Jesse H. Jones saw to it that the city capitalized on its location and economic opportunities.

West Columbia, in Brazoria County, became a big producer in 1918, the year after the discovery well came in. No doubt the rebuilders hoped Lucky Jim would live up to its name. *UT Archives*.

Oil flows into this earthen tank from pipes on the right. The four pumps above transmit the crude through a network of pipes to better storage facilities. *Exxon Corporation*.

Interior of Humble's pumping station in West Columbia. *UT Archives*.

A steam boiler furnished power to pump Richardson #1 at Barbers Hill. The enclosed boiler was located far enough away from the well to minimize the danger of sparks igniting the oil. Based in Houston, Schlueter photographed wells in a wide area. *Houston Public Library*.

To get to the top of Humble's Schilling #19 in Barbers Hill, one needed sureness of foot. The U.S. Bureau of Mines frowned on such oil-field hazards. *American Petroleum Institute.*

Forty miles northeast of Houston, the Dayton field was a modest producer in 1917. *Houston Public Library.*

Oil was struck in Hull, eleven miles south of Batson, in 1918. Amil Abel (at the well) decided to stick with his clapboard house, though surrounded by oil wells. *Houston Public Library.*

Lord of all he surveys. *Houston Public Library*.

The Pierce Junction field began with the Gulf Production Company's Taylor #2 on February 19, 1921. During that year the field produced 1,403,940 barrels. *Houston Public Library*.

Production began at Damon Mound, between Rosenberg and West Columbia, in 1915. This driller laid in a good supply of wood for his boiler. *Houston Public Library*.

The surface of this Damon Mound oil tank was as smooth as the proverbial mill pond. *Houston Public Library*.

Production began in Orange in 1913, when Rio Bravo Company's Bland #1 struck oil at 3,128 feet. This was the deepest well in the state at the time. The field surged forward again in 1921, as pictured above. *Houston Public Library*.

*Left:* Oil was not discovered at Sinton until 1934, but this gas well dates from 1916. Sinton is about twenty miles northwest of Corpus Christi. *Library of Congress. Right:* The Refugio field began as an important gas producer in 1920. Here a wild gas well is flared. *Lamar University.*

Oil was discovered at Refugio in 1928, and two years later it was the leading field on the Gulf coast. This Refugio well has an updated version of the double-deck derrick first used at Humble in 1906 (see photograph on page 84). *Lamar University.*

Other Coastal Fields / 145

This aerial view, taken in 1932, shows the great visibility of a flaming gas well on the coastal plain near Corpus Christi. *National Archives*.

In the Corpus Christi turning basin, three freighters (on the left) are moored to the wharf, while a tanker (identified by the smokestack at the rear) loads at the oil dock. Nueces Bay is in the background. To reach Corpus Christi Bay and the Gulf, the ships steamed under the drawbridge in the foreground. *National Archives.*

Houston served as the lynchpin for the Gulf coast oil business, excepting that of the Beaumont–Port Arthur area. As a metropolis, it was a market for petroleum products, such as the asphalt being spread on the street in front of the Texas Company's office in 1915. *Texas Co.*

Houston also supplied the needs of boom towns, with such items as portable jails. *Texas State Archives.*

The cruiser U.S.S. *Houston* steams past a refinery on the Houston ship channel, 1930. The electric power plant in the foreground and the refinery in the background illustrate the interrelation of heavy industry on the channel. Note the Texaco tank cars in the foreground. *National Archives*.

The Sinclair refinery on the Houston ship channel. *National Archives*.

Loading a tanker at the Shell refinery docks at Deer Park on the Houston ship channel. *Shell Oil Co.*

A Humble tank truck of the mid-1930's in Hermann Park, Houston. *Exxon Corporation.*

A fortune going up in flames. *Permian Basin Petroleum Museum.*

# Ranger

NOT every oil field or boom town in Texas had a distinguishing characteristic, but Ranger did. Water was so scarce there that men drained their automobile radiators each night lest someone steal the water.

As early as 1912 the Texas Pacific Coal Company, located in Thurber, sixteen miles east of Ranger, discovered traces of oil as it bored test holes for coal. These 700- and 800-foot holes were seven miles south of Ranger, along the Leon River. For the next several years drillers tested for oil, with only middling results. Geologists had indicated that the area had much more oil than had been tapped, and Ranger businessmen eagerly sought the means to prove them right. The businessmen had about exhausted their own resources in dry holes, and a periodic drought caused a depression among the farmers and livestock men, the backbone of the local economy. Unable to finance further drilling, some Ranger men appealed to the Texas Pacific Coal Company to undertake exploration. W. K. Gordon, the manager, was persuaded that geologists were right when they said that oil lay deeper than previous drilling had reached. He committed the company to drilling four test wells to 3,500 feet, provided the Ranger boosters could lease 10,000 to 15,000 acres. Evidencing oil fever, landowners in the Ranger community quickly offered 25,000 acres for lease, receiving twenty-five cents an acre in cash with another dollar per year for seven years.

The first well Gordon contracted for produced some gas but was abandoned. In the late summer of 1917 drillers began the second well on J. H. McClesky's farm, two miles southwest of Ranger. Gordon informed his New York office daily of the well's progress as it approached 3,000 feet. The office told him to cease drilling if no oil was discovered by 3,200 feet, but he ignored the order. On October 17, 231 feet farther, the well hit pay, flowing 1,700 barrels daily. Farmer McClesky doubtless jubilated over the good fortune of having a gusher on the forty acres he had leased, but he nonetheless criticized the careless drillers for allowing oil to splash on his white thoroughbred leghorn chickens. The company's third and fourth test wells also caused rejoicing. Not to be outdone, the first well exploded on New Year's Day, 1918, shooting millions of feet of natural gas into the air. Six weeks later, a second eruption from the well began a gusher ultimately worth half a million dollars. By the spring of 1918 Ranger's oil-field boom had swelled the town's inhabitants from one thousand to six times that number, with all the attendant excitement and problems. As exploration and production continued, the next year brought another twenty thousand people to Ranger. Among these were representatives of major oil companies, seeking leases near the producing wells. Unfortunately, from their standpoint, the Texas Pacific Coal Company (reorganized to add "and Oil" to the corporate name) seemed to have pre-empted the most likely locations. Such companies as Humble, Magnolia, and Prairie Oil and Gas solved this problem by drilling the leased land on a fifty-fifty basis with the Texas Pacific company. With the majors thus dominating exploration of the Ranger field, independent wildcatters had little chance. Another factor militating against independents was that wartime inflation had raised the costs of materials and labor. Also, drilling at depths in excess of 3,000 feet cost between $35,000 to $100,000 per well—costs out of the range of most small operators.

Resourceful independents were not eliminated entirely. Ambition and grit rewarded those who played good hunches—outside the areas controlled by the majors. One colorful wildcatter, Barney Carter, got his start in Ranger. With no oil experience or capital, he sought his fortune there after news of McClesky #1 reached his native Abilene. Soon after he arrived, he ordered a milk shake in a drugstore, only to find there was no milk in Ranger. Carter's solution to that problem was to begin a dairy, so he leased two hundred acres for grazing and returned to Abilene to get a loan to cover the cost of the lease and thirty-eight cows. As he developed his dairy, he acquired leases northwest of Ranger and drilled them profitably.

Another "innocent abroad," Cliff M. Caldwell, a young schoolteacher from nearby Breckenridge, went to Ranger with a friend, hoping to strike it rich. They stayed in one of the flimsy hotels thrown up to gouge the eager fortune seekers. Their seven-by-eight-foot room was separated from the adjoining room by a beaverboard partition. One night from the next room they clearly heard three Oklahomans talking about where they planned to spend $100,000 for leases. Before the break of day, Caldwell had mapped out the areas desired by the Oklahomans and beaten them to the draw. He patted the farmers on the back and by ten o'clock had purchased his leases at $5 an acre, which he then sold to the Oklahomans at a tremendous profit.

For the remainder of 1917, after McCleskey #1 opened the field on October 17, Ranger produced 93,053 barrels of high-grade crude. Second in quality only to Pennsylvania crude, Ranger oil had a dark greenish-black hue. It sold for $2.60 per barrel. The output in 1918 soared to 3,107,000 barrels, and with increased wartime demand the price rose to $4.25 a barrel.

From a human standpoint, the most fascinating discovery in 1918 came at Merriman, a hamlet southwest of Ranger. Sitting directly on top of the oil pool was the Merriman Baptist Church. Prospectors offered the congregation a handsome sum for drilling rights on its two-and-one-half-acre church lot, but it refused. Individual members who had previously gotten sizable royalty checks agreed to bear drilling costs. The congregation decided to donate 85 percent of its oil income to the Baptist General Convention of Texas and to keep the rest, which it would devote to a new edifice. When the congregation was approached about leasing rights in the adjacent church cemetery, they astounded the lease seeker by turning down his million-dollar offer. The pastor explained that the cemetery had been deeded to the dead and therefore the congregation had no legal right to lease it.

Economic conditions in booming Ranger made the farmers forget the debilitating drought. If they made nothing from their crops, they and their families could earn high wages in town. Unskilled clerks in stores made $60 per week; teamsters got $7 a day; roughnecks and day laborers, $12 to $15; tool dressers, $20; with drillers and rig builders drawing $25 to $30. Carpenters worked furiously to build new stores, hotels, offices, and supply houses.

Congestion was the order of the day. Oil-field workers slept in shifts in rented rooms, without changes of bed linen. Hotels utilized all possibilities for income by renting lobby chairs. The single men who thronged the town found it just as difficult to get a good meal as a decent place to sleep. Cafes could not keep enough food to supply the hungry roustabouts, and they never offered enough seating for those wanting to eat at normal meal times. Their inflated prices deterred no customers. No place in town suffered more congestion than the Texas and Pacific Railroad depot, where five trains arrived from and departed for Fort Worth daily. Space on the trains commanded such a premium that men rode on cowcatchers, on top of box cars and passenger coaches, and on the sides by hanging through open windows. Practically everything needed for the boom—groceries, lumber, dry goods, and oil-field equipment—arrived by rail. Unloaded freight cars backed up so that the railroad forbade further shipments to Ranger until the backlog was cleared. But the pressure increased and loaded cars filled sidings in nearby towns. Those who had ordered supplies found that the best way to guarantee their delivery was by tipping the appropriate railroad employees.

Like most congested boom towns, Ranger attracted a criminal element. Gamblers, thieves, whores, and bootleggers converged on any place that offered quick money, and lots of it. In one twenty-four-hour period, Ranger had five murders. The morning sun frequently found dead men sprawled in the streets. Lawless conditions in Ranger made it a natural target for the outfit from which the town took its name: the Texas Rangers. One notable raid came on Saturday night, February 12, 1922. As the casino on the third floor of the Commercial Hotel operated full blast, the Rangers pounced, catching ninety gamblers and confiscating their money and equipment. The casino's operators paid the fines of those arrested!

Since local law enforcement officers lacked zeal in prosecuting crime, the Rotary Club took it upon itself to rid Ranger of organized vice. In 1922 the Rotarians—"Service Above Self"—raided ginmills and gambling parlors, capturing $16,000 worth of liquor, which they poured into the streets. This vigilantism deterred further criminal activity in Ranger, but by then that boom was pretty well over.

At the same time as the Ranger boom, wildcatters discovered oil throughout that region—in nearby Eastland to the west and in Brownwood, fifty-five miles to the southwest. A distinctive feature of the latter field was the extremely shallow depths where oil was found. Much of the crude from this region was piped to refineries in the Fort Worth–Dallas area.

*Left:* The discovery well at Ranger, McClesky #1, October 17, 1917. (Other sources give the date as October 21 and 25.) This is one of the few scenes where spectators seem genuinely excited by a well roaring in. *Southwest Collection. Right:* Clarence R. Martin, a geologist for the Texas Company, studying the Ranger terrain, 1917. *Southwest Collection.*

Merriman lay a few miles southwest of Ranger. Although its Baptist Church was offered $1 million for drilling rights in the cemetery, the church could not entertain the offer since the cemetery was deeded to the dead. *Southwest Collection.*

These Baptists refused a substantial amount for drilling rights on their 2.5 acre lot. Since many had grown rich from royalties on their own land, they financed three wells on the church grounds. The congregation voted to give 85 percent of the oil income to the Baptist General Convention of Texas and to build a new church with the remainder. *Southwest Collection.*

This fire of April 6, 1918, leveled a good bit of boom-town Ranger. *Texas Mid-Continent Oil & Gas Association*.

A field expedient for fighting the flaming Perkins #5 in Ranger, September, 1919. The multiple boilers could provide a volume of steam under tremendous pressure to quench the fire. *Permian Basin Petroleum Museum*.

Mrs. H. H. Adams at a geologist's plane table in the Ranger field, 1919. She was the wife of a Texas Company field geologist. *Southwest Collection*.

The Ranger Police and Fire Department in the 1920's. The fire engine was obviously not intended for the oil field. *American Petroleum Institute.*

Heavy oil-field equipment, plus a little rain, turned Ranger's dirt streets into mudholes. *Texas Mid-Continent Oil & Gas Association.*

Ten miles west of Ranger lay Eastland. This well on C. U. Connellee's horse farm was one-half mile south of Eastland. With the proceeds from his oil, he began a nine-story hotel and built a theater, baseball field, and race track in Eastland. *Texas State Archives*.

The Montrose Oil Refinery, near Hodge Station, north of Fort Worth. *Fort Worth Star-Telegram.*

A front-yard well in Brownwood, 1918. The well was so shallow that an ordinary water well pump raised the oil to the surface. Profitable production began in Brownwood in 1917 at 180 feet. *Southwest Collection.*

William Earl Hubbard (*right*) a Texas Company geologist, and a friend observe a pump in the Brownwood field, 1918. *Southwest Collection.*

Originally called Hogtown, after nearby Hog Creek, the town became Desdemona when the boom began. Unlike Shakespeare's heroine, it did not die after the boom faded but reverted quickly to a quiet village. The photographer took literary license with the caption, for "spudding in" meant to start a well. *American Petroleum Institute.*

# Desdemona

ABOUT fifteen miles southeast of Ranger, denizens of Hogtown talked excitedly about oil. For years they had noticed the iridescent shimmering of oil on the surface of Hog Creek, from which the community took its name. They reasoned that if Strawn to the north, as well as other nearer locations along the Leon River, had oil, they might have enough to make testing worthwhile. Appropriately, on Ground Hog Day, 1914, some one hundred residents organized the Hog Creek Oil Company. Members of the company bought stock at ten dollars per share, raising enough to buy a cable tool rig. Their do-it-yourself exploration resulted only in a 1,500-foot dry hole.

Professionals shared the amateurs' interest in finding oil at Hogtown. In 1917 oilman R. O. Harvey of Wichita Falls dispatched William E. Wrather, a geologist, and Landon H. Cullum, a lease man, to examine the area and procure leases if they thought it had possibilities. Wrather discovered an anticline that he considered worth testing. (An anticline is an upfolded geological formation that traps oil against a non-permeable surface, such as salt. Anticlines formed when the earth's layers shifted.) Cullum rounded up leases on 6,000 acres at two dollars per acre. The Harvey group arranged for the Maples Oil and Gas Company to drill for a half interest in a 5,000-acre lease. Included in that company were Michael L. Benedum, John Kirkland, F. B. Parriott, and Joe Trees. Perhaps this evidence of professional interest in their community caused the residents to upgrade the name of their town. Whatever the reason, Hogtown had become Desdemona by the autumn of 1917.

Geologist Wrather, who in the 1940's served as director of the U.S. Geological Survey, chose the site for drilling, Joe Duke's farm. The cable tool crew of Pete Hoffman went to work and on the night of September 2, 1918, brought in a roaring gas well. As the gas spread through the well site, a spark from the tool dresser's forge ignited it, illuminating the whole farm. For three days the well blazed, until R. O. Harvey brought it under control with a stream of high-pressure steam produced by four boilers. By the end of the month Hoffman had struck oil at 2,960 feet, bringing in

2,000 barrels daily. A month later this Duke #1 produced 3,840 barrels a day. The next well drilled was on the Knowles farm, and it flowed 8,000 barrels daily.

The Desdemona boom had begun! Unlike the Ranger boom, where the Texas Pacific Coal and Oil Company controlled much of the leased acreage, in Desdemona leases were mostly small. One was for one-hundredth of an acre. As Desdemona proved to be a rich field, the usual personnel converged in droves, the village's population quickly swelling to 20,000. By January, 1919, hundreds of drilling crews were sinking wells, with some spectacular results. The Payne gasser, 400 yards from Duke #1, pumped 30 to 40 million cubic feet of gas into the atmosphere with such pressure that the roar could be heard twenty-five miles away. The field's richest producer, the Hogg well, a mile northwest of Duke #1, flowed 15,000 barrels daily.

Storing and transporting the field's heavy output strained existing facilities and necessitated many makeshift ones. Besides digging sumps for oil, operators dammed creeks and gullies, which filled quickly. Since Desdemona was not on a railroad, crude oil was originally hauled to nearby railroad towns in tank wagons. Then producers laid pipelines to Gorman and DeLeon, which had a railroad. Magnolia and Humble participated in the boom, with the former handling a major share of storing, transporting, and refining Dedemona's crude.

Desdemona held the spotlight only briefly. Its peak year was 1919, when it produced 7,375,825 barrels. The next year it slumped to 2,767,115. Wasteful practices, common to the industry, were responsible. Operators gave gassers their heads, hoping they would perform as Duke #1 had and turn into gushers. They let gushers flow freely to advertise the field, trying to attract more investors. These practices dissipated the underground gas pressure that could keep crude flowing to the surface. When natural pressure failed, wells were put on pumps. But pumps could not guarantee maximum recovery, since, with the exhaustion of gas pressure, much oil receded from a pool that could be pumped. C. C. Rister reported that only one-half the field's potential was ever realized.

Today if one travels the back roads to Desdemona—the only way to get there—he finds a village of about a hundred inhabitants and scarcely any vestiges of the roaring boom town of 1919. In town there is a concrete foundation for a derrick, grown over with weeds. On the outskirts, a few wells pump lazily, under the faded sign of the flying red horse, the symbol Mobil inherited from Magnolia.

*Left:* A spark from the tool dresser's forge ignited Desdemona's discovery well on Joe Duke's farm, fifteen miles southeast of Ranger. On September 5, 1918, R. O. Harvey of Wichita Falls quenched the fire, which had burned three days, with steam from multiple boilers, as in the photograph on page 158. *Southwest Collection. Right:* Drillers prepare to "shoot" a well in Desdemona. A "torpedo" of nitroglycerine was detonated at the bottom of the well to release oil from the surrounding porous rock formations. *Southwest Collection.*

In 1910 this well shooter hauled his nitroglycerine torpedoes in a buckboard. *Shell Oil Company.*

Desdemona  /  167

Before pipelines could be laid to transport Desdemona's oil to refineries, local storage had to be provided. Earthen tanks, such as the above holding 200,000 barrels, were the cheapest and quickest to construct but also the most susceptible to fire, contamination, and evaporation. *Southwest Collection.*

A panorama of the Desdemona field taken from the top of a storage tank, 1919. *Library of Congress.*

Although the Desdemona field extended from southeast Eastland County into Erath and Comanche counties, it was also right downtown. The pyramidal tent in the lower right corner might well have come from a World War I surplus store. *Permian Basin Petroleum Museum.*

Downtown Desdemona, where parking space commanded a premium. *Texas Mid-Continent Oil & Gas Association.*

Since Desdemona couldn't begin to accommodate the thousands of workers who had poured into the town by 1920, most had to live in tents. *American Petroleum Institute.*

Fast-buck artists descended on boom towns almost as quickly as roustabouts, and money circulated rapidly. *Southwest Collection*.

"Giving Her The Soop"

Drillers lowering a nitroglycerine torpedo into a Breckenridge well in 1922. Note the length of the torpedo on the left. The pipe to the right of the hole shows that the flow would be directed immediately into a tank. *Southwest Collection.*

# Breckenridge

SHARING in the region's boom-town excitement of 1918 was Breckenridge, county seat of Stephens County, some twenty-five miles northwest of Ranger. The first exploration in the county had come in 1911, but no oil was found. In 1916 drillers brought in a modest well near Caddo, fourteen miles east of Breckenridge. For the next few years, the Caddo field produced enough oil to encourage further exploration, especially around the county seat. Before oil was struck there, Breckenridge boasted a summer population of around 600. During the school year the number might climb to 1,000 as farmers and ranchers moved their families to town so that children could get schooling. With the boom in full swing in 1919, Breckenridge's population had reached 30,000.

The first significant indication of oil there came in 1916, but two years later when Smith #1 blew in as a gas well, then began spraying the countryside with oil, it set off the typical scramble for leases, with soaring royalty prices and an influx of hopefuls. J. Chilton Lynn was one of those. He was earning fifty dollars per month as an automobile mechanic in San Angelo but migrated to Breckenridge when he got a similar job there paying fifty dollars per week. He then transferred his mechanical ability to the oil field to become a successful wildcatter. Within the city limits, the first important well was Stoker #1. Chaney #1 eclipsed it, however, on February 4, 1918, when it proved a huge producer. Almost overnight, two hundred derricks sprouted in town.

Breckenridge differed from many neighboring towns by allowing drilling within the city limits. Its property owners banded together by blocks to share royalties on any oil found under the block, thus obviating frantic drilling on every available foot of ground. Since the town sat right over a huge oil pool, a booster pointed out that all wells drilled in town were fine producers. Peak production of 31,037,710 barrels came in 1921, with a rapid decline thereafter. In 1925, only 5,729,000 barrels came from the field. With the decline of the field, Breckenridge's economy tumbled and many banks closed. As a monument to what the field had been, 2,000 derricks still rose from city lots.

While the field flourished, it brought the usual social chaos to the community. The Stephens County jail bulged with two hundred prisoners per month, virtually all associated with the oil field. They included murderers, hijackers, bootleggers, and gamblers. Whores were so brazen that they solicited at the derricks. Their working on location depended on the progress of the well and the scruples of the driller. Hoodlums playing the protection racket also visited the field, trying to shake down the roustabouts. Those with no money were supposed to get some and leave it the following day at a specified place, under the threat of bodily harm. The American Legion stepped in to control the criminal element, and their efforts were supplemented by those of the Texas Rangers.

By 1925 all signs indicated that the field was failing, and many businessmen connected with the boom had already folded. For the next seven years it limped along with decreasing production. Then in 1932 Breckenridge became the site of an important experiment called acidization. Hydrochloric acid was poured into wells to see if it could restore production by breaking down limestone particles to release petroleum into a pool that could be pumped. The experiment worked and Breckenridge revived. The Chemical Process Company charged $360 to treat a well with 2,000 to 3,000 gallons of hydrochloric acid, and the well's output was usually doubled thereby. Building on its success at Breckenridge, the Chemical Process Company revitalized fields in Louisiana and Oklahoma, as well as Texas.

K. STOKER Nº 2.
12,000 BBL. GUSHER
LARGEST IN STEPHENS Co.
BRECKENRIDGE, TEXAS.
Gulf Production, Co.
Photo By
~ The Texas Studio ~

Oil was discovered in Breckenridge, some 25 miles northwest of Ranger, in 1916, but the boom didn't start until 1918. Drilling in town began with Stoker #1, but #2 heightened the frenzy. *Permian Basin Petroleum Museum.*

Stoker #2 quickly filled the metal storage tanks and the overflow went into the sump, thoughtfully labeled for the viewer by the Texas Studio. *Permian Basin Petroleum Museum.*

As in Beaumont and Wichita Falls, Breckenridge had its sidewalk stock exchange where brokers with get-rich-quick schemes peddled shares in new oil companies. *American Petroleum Institute.*

At the height of its boom in 1921, Breckenridge, county seat of Stephens County, boasted a population of 30,000. During that year it produced 31,037,710 barrels of oil. *Permain Basin Petroleum Museum.*

A traffic jam on East Walker Street in Breckenridge. *American Petroleum Institute.*

On the main street of Breckenridge in 1920 oil-field workers could see *Sign of Jack O'Lantern* and *The Four Seasons* at the National Theatre. *American Petroleum Institute.*

The clapboard shacks in the foreground were built to accommodate the influx of oil-field workers into Breckenridge. Note the privies to the rear of the dwellings, as well as the pipes omnipresent in the oil towns. *Exxon Corporation.*

John Seratch, the teaming contractor, offered to haul oil-field equipment anywhere in Breckenridge. *American Petroleum Institute.*

Oil company pipeliners turning their talents to civic duty. *American Petroleum Institute.*

The Mexia discovery well, L. W. Rogers #1, November 19, 1920, brought in by Colonel Albert E. Humphreys. After the exciting gushing, the well averaged a modest fifty barrels daily. *Exxon Corporation.*

# The Mexia Fault

ALTHOUGH the emergence of Mexia as a vital oil field dates from November, 1920, for eight years the area had served as a major gas field, thereby giving some indication of what might develop. Blake Smith, a local businessman, organized the Mexia Oil and Gas Company in 1912 to explore for gas in the area. In forming the company he had persuaded one hundred Mexia businessmen, about all there were in the town of some four thousand, to pledge $1,000 each to drill ten wells. When the first nine produced only traces of gas, Smith appealed to the drilling contractor's gambling instincts and got him to drill another two for the price of what had been intended as the tenth and last. The tenth well was also a failure, but the eleventh justified Smith's faith. It became the cornerstone of the Mexia gas industry, supplying all Mexia's fuel needs, plus those of Waco, Corsicana, Mart, Groesbeck, and Teague, all within a forty-mile radius.

For several years the gas field prospered, producing well over two billion cubic feet in 1915. But like other gas fields before it, such as Corsicana and Petrolia, it began to slump, finally playing out at the height of Mexia's oil boom and creating an acute fuel shortage for the 40,000 who had swarmed into the town. Smith and his associates in the Mexia Oil and Gas Company were not distraught over the failure of their field, for they firmly believed that it also contained oil. They offered half interest in their 2,000-acre lease to any operator who would drill for oil.

The way Mexia's discovery well got completed illustrates the complications that often attended oil exploration. Smith first approached Colonel Albert E. Humphreys, successful head of the Homaokla Oil Company, but he was uninterested. His consulting engineer, J. Julius Fohs, however, thought Mexia worth investigating and got W. A. Reiter and John A. Shephard to join him in the enterprise, the latter contracting for the drilling. They selected the L. W. Rogers farm, three miles west of Mexia, as the site of the well and spudded in during September, 1919. The driller lost the hole at 109 feet, so he transported the derrick a short distance away and began again. This new hole went poorly, exhausting Shephard's funds. Thereupon, he offered half interest in the well to the highest bidder. Colonel Humphreys had kept his eye on the operations and decided it was worth a gamble. When Rogers #1 was finally completed at 3,100 feet on November 19, 1920, Humphreys' faith was justified—but not emphatically. The well produced only fifty barrels daily, but that was enough to convince Humphreys of the need for further exploration.

Humphreys had gambled not only on the Rogers well but on the entire area. His daring showed how the audacious wildcatter could make a quick fortune in Texas oil. Even before Rogers #1 was completed, the cuttings had shown traces of oil. Acting on this cue, Humphreys began building a 1,600-barrel tank and negotiated with the Texas Company to lay a pipeline to the field and to build a loading rack at the railroad in town. Then he and Fohs formed two new outfits, the Humphreys-Texas and Humphreys-Mexia companies, and proceeded to lease 12,000 acres lying along the Mexia fault, where the Rogers farm was situated. With the completion of Rogers #1, Fohs advertised that the companies would "prove up the field," making an effort "to conserve as much of the oil as possible." Obviously, by 1920 the futility of the immense waste connected with earlier Texas oil explorations had been driven home. Operators had come to realize that the spectacle of gushing, however exciting, ultimately reduced productivity, hence profits, and should therefore be curbed.

The Humphreys companies set about systematically to fulfill their goals. They drilled eight wells between Mexia and Groesbeck, bought a 352-acre plot two miles south of Mexia for a refinery, and contracted for a dam to create a 135-acre lake whose water the refinery would use. Humphreys' second well, Berthelson #1, substantiated his optimism in the field. When it began flowing 4,000 barrels daily in the summer of 1921, the Mexia boom began in earnest.

A number of companies came into the field and many good wells developed, including the Blake Smith #1, Liles #1, and Henry #1. The real sensation came on August 21, 1921. The Western Oil Corporation's Desenberg #1, from a depth of 3,059 feet, began spewing forth 18,000 barrels daily, and the Adamson #1 surpassed that with 24,000 barrels per day. The latter continued to be the premier well along the Mexia fault. Little wonder that all roads led to Mexia and that they were jammed!

The area of heaviest production was called the Golden Lane, which was about one-half mile wide and ran right along the fault. In 1921 the field yielded 5 million barrels, with the number increasing in 1922 to 35.12 million. The peak day's production of 176,000 barrels came on February 12, 1922.

Anticipating the inevitable tapering off in the Mexia field, Humphreys and Fohs extended their explorations along the fault. Between Mexia and Corsicana, they brought in the Currie field on November 14, 1921. Although profitable, the Currie output was never sizable. Its best yield was 13,000 barrels a day from twenty-two wells. The partners also tested around Kosse, twenty-eight miles south of Mexia. Their discovery well flowed handsomely for two days and stopped. Other wells there were dusters. While that Kosse venture proved costly, later drilling was profitable. As in the Mexia field proper, other companies explored along the fault. Just north of the Currie field, the Boyd Oil Company struck oil at Wortham on November 22, 1924. The next year Wortham produced 16,838,150 barrels from three hundred wells.

Far to the south, where the Mexia fault crosses the San Marcos River, the town of Luling is situated, forty-five miles south of Austin. In the early 1920's Luling and surrounding Caldwell County enjoyed speculative leasing. Among those hoping to find oil there was Oscar Davis, who bought $75,000 worth of stock in the Texas Oil and Lease Syndicate. He proposed that his brother Edgar B. go to Caldwell County to oversee the lease, promising him one-third of the profits. At the time Edgar B. Davis had retired from a successful career in the rubber business. He had been responsible for the United States Rubber Company's planting its first five million plants in Sumatra.

Edgar decided to accept his brother's invitation and journeyed to Luling, which he found far different from Sumatra or his more recent residence in New York. Board sidewalks and unpaved streets suggested little economic enterprise, but Davis nevertheless looked on the area as a challenge. Lack of previous experience in petroleum did not deter him from operating in the grand manner. He mortgaged himself to the hilt to buy out the interests of his brother and others in the Texas Oil and Lease Syndicate and formed a new outfit, the United North and South Oil Company.

After drilling three dry holes, the company had a lucky strike with its Rios #1 on August 9, 1922. From 2,175 feet the well produced 150 barrels a day, but by that time Davis was broke and unable to meet his payroll. Providentially, his secretary found at his summer home in Massachusetts several securities he had forgotten. By selling those, he could keep the company afloat. When Rios #2, plus other good wells, came in, the immediate financial crisis had passed.

When Rios #1 established the possibilities of the Luling field, representatives from Magnolia visited Davis. They contracted to purchase one million barrels of crude for fifty cents per barrel. Moreover, they constructed a pipeline from the field and lent Davis' company $300,000. This assistance, in addition to the income from the contract, enabled Davis to proceed in developing the field. His efforts were sufficiently fruitful that on June 11, 1926, Magnolia bought him out, paying $12.1 million, divided evenly between cash and oil. Magnolia's purchase consisted of 215 wells pumping 17,000 barrels daily. Davis' experience documented the rewards that could follow a bold gesture. Since he entered the oil business a complete neophyte, it was fortunate that the bold gesture was also coupled with considerable luck.

*Left*: At the height of Mexia's boom in the early 1920's, General John J. Pershing spuds in the J. K. Hughes #1. Immediately behind him, the big man is Colonel A. E. Humphreys, who invited Pershing to Mexia. To Pershing's right is Pete Urschel, drilling superintendent for Humphreys. Earlier Hughes had drilled two producers on the Mills property that he sold to Magnolia. He drilled twelve others on Ed Ellis' fifty acres. *Southwest Collection*. *Right*: Another of Colonel Humphreys' successful ventures, 1921. *American Petroleum Institute*.

Approximately one-half mile wide, the Golden Lane was Mexia's most productive area. It lay right along the Mexia fault. In 1922 the Golden Lane yielded 35 million barrels. *Exxon Corporation*.

When news of the Mexia boom got out in 1921, the town's population leapt from 4,000 to 40,000 within days. *American Petroleum Institute*.

The flooded Mexia street presented less of a problem to the team than to the Model T's. *Texas Mid-Continent Oil & Gas Association*.

Twenty-eight miles south of Mexia, the Kosse field proved erratic. This well, which gushed on August 18, 1922, produced thousands of barrels for two days, then sanded up. When cleared, the well yielded 9,000 more barrels (on September 13), but the flow lasted only thirty hours. After being drilled two feet deeper, it produced by fits and starts but was economically insignificant. Note the pipe fittings in the foreground. *Library of Congress.*

Edgar B. Davis' United North and South Oil Company brought in the Rios #1 on August 9, 1922. In 1926 Magnolia paid Davis $12.1 million for his Luling holdings. Luling is fifty-seven miles east of San Antonio, where the San Marcos River crosses the Mexia fault. *Shell Oil Co.*

This name furnishes an understandable variation on the "land and cattle company" combination famous in West Texas. *San Angelo Standard-Times.*

# Big Lake

THE strike that heralded the Permian Basin as a stupendous source of oil came in the Big Lake field on May 28, 1923. This vast arid basin, once an inland sea, covers 76,610 square miles in Texas, plus 12,000 in southeastern New Mexico. As part of its program for supporting public education, the Republic of Texas had devoted millions of acres in the Permian Basin to the schools. When the University of Texas was created in 1883, the land was assigned to it, with one-third of any income earmarked for the Agricultural and Mechanical College of Texas, now Texas A&M University. At the time, the desert land was practically worthless. Its huge ranches supported few cattle, simply because of scarce water and vegetation. Because the land was so unproductive, the university happily realized a little income from those wishing to explore for oil and gas.

Among those searching for petroleum in the area was the Underwriters Producing Company. Early in 1920, Steve Owen, the company's field manager, began a well near Westbrook, ten miles west of Colorado City. The well, T. and P.–Abrams #1, began flowing in June, but only twenty-five to thirty barrels a day. As the curtain-raiser for the prolific Permian Basin, this well gave puny indication of what lay ahead. But it did stimulate exploration throughout the region—as far west as Van Horn, which remained dry.

The first to plan systematically to search for oil in Reagan County was Big Lake lawyer Rupert P. Ricker. In 1919 he filed drilling permits in Reagan and surrounding counties but could not raise the ten-cent per acre rental. He succeeded in getting an army friend, Frank T. Pickrell, to take the lease. With fellow El Pasoan Haymon Krupp, they leased 431,360 acres in Upton, Crockett, Reagan, and Irion counties and formed the Texon Oil and Land Company. Their plan was to drill in southwest Reagan County, but the lease specified that a well had to be begun by a certain date. Because of difficulties in raising money for drilling, the prospectors had to improvise. The town closest to the intended well site was Best, nine miles west of Big Lake, so Pickrell and Krupp had the drilling machine shipped there. All they could afford was a water well driller, but that would at least begin the hole and they could comply with their lease provisions. Pickrell loaded the machine on a wagon, proceeded four miles west to where he had staked the well, and then began drilling. According to their state permit, they had to have two impartial witnesses sign affidavits that they had seen the well begun. With their time running out, they feared they could find no one handily. Appearing as if from a mirage, two men drove up in an automobile and agreed to witness the spudding.

Even before that risky beginning, associates of Pickrell's in New York suggested that prospecting for oil in the Texas desert smacked of madness. They told Pickrell he should invoke the aid of the patron of the impossible, Santa Rita, by climbing to the top of the derrick and showering petals of a rose that had been blessed on the spudding machinery below. He followed their advice, christening the well Santa Rita, as the drilling began on September 3, 1921. What the two witnesses thought of that was not recorded in the affidavit. Almost two grueling years later, on May 28, 1923, the driller, Carl Cromwell, struck oil at 3,028 feet and Santa Rita #1 opened a new era for the economy of West Texas.

The location of the well compounded problems for the developers. Because it was far from any existing fields, equipment was difficult to get. Even after oil was struck, Pickrell had to wait a month for the arrival of pipe with which to direct the oil into storage tanks. Then he had to negotiate for pipelines to transport the oil out of the field. One of the first companies he dealt with was Benedum and Trees, of Pittsburgh. These dealings eventuated in the formation of the Big Lake Oil Company. Michael Benedum, "the great wildcatter," was no stranger to Texas oil, having participated in the development of Desdemona. Holding 75 percent of the Big Lake Oil Company's stock, Benedum agreed to drill eight additional test wells. Other firms Pickrell involved in developing the Big Lake field were the Marland Oil Company and Humble. The latter built an eight-inch pipeline, carrying 20,000 barrels daily, from the field to its Comyn station in the Ranger area. From there, the oil was pumped to the Baytown refinery. The first Big Lake oil flowed through the line on April 18, 1925.

By then, the Big Lake field had seventeen wells daily yielding 11,500 barrels from a shallow formation around 3,000 feet. The field actually was adjacent to the little company town of Texon, thirteen miles west of Big Lake. The field reached its peak on August 31, 1925, producing more than 40,000 barrels. At the end of the year, it had yielded 10,060,330 barrels. The University of Texas grew unexpectedly rich from its Permian Basin oil income, all beginning with Santa Rita #1, which is pumping yet.

A new era began when Santa Rita #1 blew in on May 28, 1923, tapping the vast wealth of the Permian Basin. Named for the patron saint of the impossible, this still-producing discovery well of the Big Lake field is about seventy miles southwest of San Angelo. The field is actually at the tiny town of Texon. *Texas Mid-Continent Oil & Gas Association.*

Despite the boosterism of the 1922 sign, Van Horn never made it as an oil center. The great discoveries of the Permian Basin lay to the east and north. *Humanities Research Center, UT.*

To celebrate the well's contribution to the University of Texas, these portions of the original rig became an outdoor museum on the Austin campus. When Walter Prescott Webb was president of the American Historical Association in 1958, he convened the council with a gavel carved from the Santa Rita rig. *Permian Basin Petroleum Museum.*

This rig replaced the original at Santa Rita #1 after the latter was removed to the university campus in Austin. *San Angelo Standard-Times.*

*Left:* A Texon gasser. *Library of Congress. Right:* Wise men from the East, Levi Smith (*left*) and Michael L. Benedum, at the Big Lake field, 1925. Benedum, one of the nation's most illustrious oil men, participated in developing this, as well as the Desdemona and Yates fields. Levi Smith managed the Big Lake field. Today on Ranch Road 1555 in barren southeast Upton County, the Sarah Benedum camp houses workers at Phillips Petroleum's Benedum plant, a few miles to the east. *San Angelo Standard-Times.*

The Big Lake field. *San Angelo Standard-Times.*

Because of Texon's isolation, the company provided proper schooling for its children. Note the pipes at the right of the building. *Texas State Archives*.

The camp for Big Lake oil-field workers. *Permian Basin Petroleum Museum*.

On October 3, 1925 George B. McCamey brought in this discovery well in the McCamey field, about fifty miles south of Odessa. The wooden rig used in drilling was converted into the pumping unit. *Shell Oil Co.*

# The Permian Basin

THE fabulous Big Lake strike spurred drilling the length and breadth of the Permian Basin. In October, 1925, George B. McCamey discovered in the southwest corner of Upton County the field that bears his name. Although that field never yielded the great quantities of the later Hendrick and Yates discoveries, it was nevertheless highly profitable. In the quest for profits, wildcatters searched for oil-bearing reefs in all parts of the basin, including areas near Big Spring, which were rewarding, and San Angelo, which were less so. But San Angelo developed into a regional center for the oil business, its banks furnishing capital and its warehouses, supplies.

Nineteen twenty-six demonstrated that the Big Lake field had only offered a sample of the fabulous petroleum wealth of the Permian Basin. In that year two great new fields emerged, the Hendrick pool of Winkler County and the Yates pool of Pecos County. On the ranch of T. G. Hendrick near Kermit, forty miles due west of Odessa, R. A. Westbrook and Company developed the discovery well in mid-July. From 2,525 feet, it produced 30 barrels daily, but the driller was dissatisfied with that yield and kept working. When he hit 3,049 feet on March 28, 1927, he got a flow of 400 barrels. Only a year later the Hendrick field was the most important in the Permian Basin, with a potential daily production of 2.75 million barrels. Because of the Texas Railroad Commission's proration, the field could yield only 150,000 to 175,000 barrels a day after July, 1928.

Soon, however, the field on Ira G. Yates' ranch would surpass the Hendrick field. The Mid-Kansas and Transcontinental Oil Company began drilling Yates #1 in a canyon two miles south of the Pecos River and thirty miles south of Rankin on October 5. The site was a dome formed by a clearly visible anticline. The well was completed on October 28, 1926, at 997 feet, producing 135 barrels per hour. Like the Hendrick discovery well, Yates #1 was successively deepened until it hit 1,150 feet, where the flow was 2,950 barrels an hour. The flow of 70,824 barrels daily on August 18, 1928, put Yates #1 in a class with the Lucas gusher.

As with other major strikes, that on the Yates ranch quickly attracted a crowd. Besides the curious onlookers who came to gawk at the gusher, men eager to lease other land on the ranch appeared on October 29. During the day Yates sold approximately $180,000 in leases. For those who wanted to stay overnight, the only accommodation was Yates' big red barn. He quickly renovated it to serve as a rooming house. With the life of the field assured, a town developed nearby, Iraan, named for Yates and his wife and pronounced Ira-An.

The Transcontinental Company continued drilling successful wells, and by mid-year 1928 it had 207 wells that put out 2.5 million barrels daily. A year later 306 Yates wells were capable of yielding 4.5 million barrels a day, but the Texas Railroad Commission had prorated production to 130,000 barrels daily.

With the prolific production from Permian Basin fields, prorationing became the only sensible solution to the problem of production far beyond refining capacity or market needs. The Humble company built a pipeline to connect the Yates field with its refinery at Ingleside, just north of Corpus Christi across the bay. Its 30,000-barrel daily capacity could handle only a fraction of the Yates production, however. The president of Humble, W. S. Farish, asked Yates producers to come to Houston in August, 1927, to discuss the problem. He urged them to agree to leave in the ground oil that could not be marketed, rather than follow their natural inclinations of producing maximally and then storing the oil. Because of the high sulphur content of Permian Basin oil, it quickly corroded metal storage tanks. By keeping oil underground, it could be conserved until needed, with no wastage or needless production expenses resulting. A month later producers met in Fort Worth and subscribed to the kind of voluntary prorationing Farish had suggested. The Texas Railroad Commission assumed responsibility for administering prorationing on July 1, 1928.

Olin Bell's Model T Ford on a geological field trip in Southwest Texas in 1925. The left foreground clearly shows an exposed fault line. *Exxon Corporation.*

In 1927 an expectant crowd gathers at another well in the McCamey field. *San Angelo Standard-Times.*

*Left:* An alert photographer captured this dramatic conflagration of the Skelly-Amerada University #1. The crew seems eager to put some distance between them and the flaming well. *Southwest Collection. Right:* Roberts' well, Big Spring. *Humanities Research Center, UT.*

Drilling in the Big Spring field. *Humanities Research Center, UT.*

Oil brought excitement to the wide open spaces of the Permian Basin—witness the crowd meeting the train at San Angelo. *San Angelo Standard-Times.*

The elevation of tank 504 enabled these men to view a considerable portion of the Southwest Texas plains. *San Angelo Standard-Times.*

Winkler County's first producer, the T. G. Hendrick #1, was brought in by R. A. Westbrook and Company in mid-July, 1926. *Fort Worth Star-Telegram*.

Wildcatting in the sand near Kermit, county seat of Winkler County, forty miles west of Odessa. *Southwest Collection.*

Hendrick B-1 with a grasshopper counterweight, 1930. *Permian Basin Petroleum Museum.*

On the Hendrick Lease Sec. 45, between Kermit and Wink, these tanks separated salt water from crude oil. The flow tank on the right received the well's production, pumped the crude into the settling tank on the left, and eliminated the salt water. In the 1920's the salt water was dumped onto the ground, devastating the sparse desert vegetation. More recently, oil producers have been required to inject the salt water deep into the earth. *Permian Basin Petroleum Museum.*

Snow from the winter of 1928–1929 softens the lines of the American Republic Corporation's T. G. Hendrick #2 on the highway from Kermit to Wink. The well still flows. *Permian Basin Petroleum Museum.*

While Winkler County was rich in oil in 1927, its highways remained primitive. *American Petroleum Institute.*

The Yates field, on the eastern border of Pecos County, was "Queen of the Pecos." The Mid-Kansas and Transcontinental Oil Company discovered oil on the Ira G. Yates ranch on October 28, 1926. The field was the finest of the Permian Basin. This Yates #2 well was 537 feet due west of the discovery well. *American Petroleum Institute.*

Yates Camp #1, 1928. *American Petroleum Institute.*

Storage tanks in the Yates field are dwarfed by the vast sweep of terrain in Pecos County. *San Angelo Standard-Times*.

Even in 1927 drillers continued to sink wells so close together that a man could walk from one end of the field to the other without touching ground. The cowboy on the left disdained walking anyhow. *American Petroleum Institute*.

The first anniversary parade in Borger, March 8, 1927. At that time the town consisted of a two-mile-long main street. *Panhandle-Plains Historical Museum*.

# The Panhandle

THROUGH the scholarly research of Professor Charles N. Gould, the vast Panhandle field of gas and oil was discovered. A geologist at the University of Oklahoma, Gould first noticed indications of petroleum in the area in 1905, but not until 1916 did he have occasion to apply his research in a commercial venture. That application resulted in the Panhandle's first well, Masterson #1, which blew in as a gasser on December 13, 1918. That well was located approximately thirty miles north of Amarillo. Soon after the initial discovery, Masterson #4 began emitting 107 million cubic feet of gas daily from 1,670 feet, and the field was immediately acknowledged as a major gas producer, creating a boom throughout the Panhandle. Landowners all across those vast plains sought Gould's professional opinion on the possibilities of their property. By the time the field was fully explored, its dimensions were staggering. With an average width of 20 miles, it stretched 115 miles, covering parts of Moore, Potter, Hutchinson, Carson, Gray, and Wheeler counties. This easily constituted the world's largest gas field.

But gas was not enough, and gas usually indicated the presence of oil, so the search continued. The Gulf Production Company earned credit for the Panhandle's first oil well with its Burnett #2, completed on May 2, 1921. From 3,052 feet it produced 175 barrels daily of a waxy crude that congealed at 50 degrees Fahrenheit, so the oil had little commercial value. Not until 1926 did the Panhandle's oil boom begin.

The Dixon Creek Oil and Refining Company opened what was to become the Borger field with its Smith #1 on January 11, 1926. This well began flowing 10,000 barrels daily, a most propitious indication of what was to come. Geologists from major oil companies converged on the site, located near the Canadian River in southern Hutchinson County, forty miles northeast of Amarillo. From Oklahoma came representatives of the Phillips and Marland companies. An enterprising real estate promoter from Tulsa, A. P. ("Ace") Borger, decided that since the site looked destined to become a major oil field, it should have a town to go with it. He secured a 240-acre tract in February, 1926, and named it for himself. His expectations in the field's future were entirely justified. By September it had 813 producing wells, with a daily output of 165,000 barrels. With wells venting their "waste" gas, as in most other fields, an odor of sour gas fouled the air.

As Borger developed a population of 10,000, it had one long main street, consisting of the usual sequence of cafes, dance halls, hotels, bars, gambling parlors, barber shops, filling stations, and dry goods stores. The barrenness of the terrain made the cheap, hastily constructed buildings look even flimsier. Social conditions there were as wild as in any previous boom town, perhaps wilder since the entire community was new. No local roots or traditions existed to temper the ordinary lawlessness of a boom town.

Gambling, whoring, and bootlegging constituted the routine crimes, with murder, assault, and robbery sounding a more ominous note. Ace Borger himself fell victim to his town's lawlessness, being shot after an argument in the post office building. The town's first jail was a chain fastened between two trees. Prisoners wore leg irons, the chains of which were then locked to the suspended chain. Lawrence Hagy, a wildcatter in the field, recalled the flourishing "jake joints," saloons serving Jamaican ginger, a crude alcoholic drink popular during Prohibition. Those who imbibed it frequently, however, found themselves unfit for oil-field work, since it produced "jake leg," or paralysis of the feet and legs. Because of Borger's lawlessness, the Texas Rangers frequently raided as many of its illegal operations as it could locate. Hagy said that after these raids, disheveled whores, carrying cheap suitcases and douche bags, lined the road, trying to hitchhike into Amarillo.

The development of Borger had a tremendous impact on Amarillo and the entire economy of the area. In 1920 Amarillo had 15,494 inhabitants, but by January, 1927, 53,000 were estimated to be there. From 1925 to 1926, assessed property values leapt from about $29,000,000 to $40,000,000, and bank deposits soared from $10,197,364 to $24,721,782. Amarillo clearly operated as the financial headquarters for the Panhandle's oil boom.

Because of Borger's isolation, transportation facilities did not exist at the time of the strike. The nearest railroad, the Santa Fe, was in the small town of Panhandle, twenty-three miles to the south. Receiving all the freight for the Borger field, Panhandle recorded a tonnage in 1926 second only to Chicago. In an effort to participate in the boom, the Rock Island Railroad began building a track from Amarillo to Borger. Santa Fe decided to extend its line from Panhandle into Borger, and the two railroads began to race. Since the Santa Fe had a headstart, it won, at a cost of $4.5 million.

Of the companies that speculated in the Borger field, one decided to stay, the Phillips Petroleum Company. It built a major refinery next to the field, lending stability to the area. Phillips enabled Borger to make the transition from a wide-open boom town to a respectable community. Later Borger acknowledged its indebtedness to the company by naming its junior college after Frank Phillips, the Oklahoma barber who parlayed a foreclosed mortgage on a filling station into one of the nation's major integrated oil companies.

The Borger field, forty miles northeast of Amarillo, opened on January 11, 1926, when the Dixon Creek Oil and Refining Company's Smith #1 came in, producing 10,000 barrels a day. The following month A. P. Borger, a townsite promoter from Tulsa, bought 240 acres and founded the town, to which he modestly gave his name. This photograph shows the field on both sides of the Canadian River. *Panhandle-Plains Historical Museum.*

A little rain rendered Borger's main street impassable. The pipes spilled from the truck in the center didn't improve matters. *Borger News-Herald.*

This unusual enclosed derrick in Borger was photographed on October 7, 1926. *Fort Worth Star-Telegram*.

The Marland Oil Company built this camp outside Borger to house its workers and their families. *Panhandle-Plains Historical Museum.*

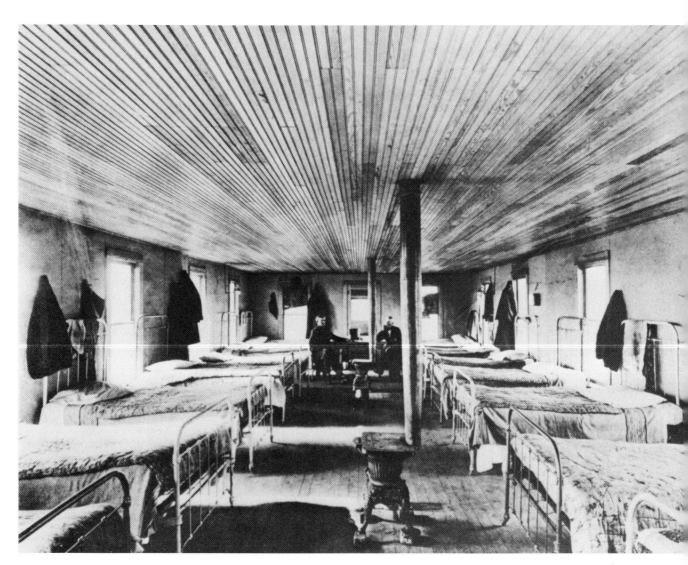

A bunkhouse for oil workers without families. *American Petroleum Institute.*

Phillips Petroleum, a giant from Oklahoma, quickly moved into the Texas Panhandle field. The Phillips camp, three miles northeast of Borger, encompassed both production and refining in the late 1920's. Today a modern refinery stands on this spot in the town of Phillips. *David Cochran, Amarillo.*

The office at Phillips' first refinery in the camp above. *Phillips Petroleum Company.*

Tex Thornton was the Panhandle's premier "nitro shooter." A woman transports his torpedoes to a well near Borger. *UT Archives.*

Although they "bored" for oil at Adrian in the 1920's, discovery wasn't until 1957. Adrian is forty-six miles west of Amarillo. *Panhandle-Plains Historical Museum.*

Most of the Panhandle drilling was by the cable tool method. Here a worker inserts a bucket to collect cuttings from a hole being drilled by this percussion method. *Southwest Collection.*

In addition to transporting oil by rail (*right background*), the Pan-Tex Company piped oil out of the giant Panhandle field. *American Petroleum Institute.*

This October 16, 1926, view is south-southwest from the Pan-Tex Camp radio tower. In this field the Santa Fe Railroad had thirty miles of main-line tracks, as well as thirty miles of sidetracks for tank car storage. *American Petroleum Institute.*

# Van

AMONG the important oil fields of Texas, Van has not had much popular notice. Located chronologically between the Panhandle giant and the East Texas behemoth, it easily escaped general attention. Van merits notice, nevertheless, because of important developments there. Twenty miles northwest of Tyler, Van lies along the eastern boundary of Van Zandt County. Before the oil boom began in 1929, Van was a sleepy hamlet surrounded by small farms.

The Pure Oil Company first exhibited an interest in Van, employing a seismograph crew to take soundings in the area in the summer of 1927. The seismograph is a machine that records impulses from small explosions set off in the earth. When the sound waves from the explosion hit oil-bearing formations, they return to the surface in configurations that indicate the presence of petroleum. The Van field was one of the first to be discovered in this manner. The highly favorable reports from the seismograph crew induced Pure to buy extensive leases in the area and to proceed to develop the field. On October 14, 1929, Pure brought in Jarman #1, producing 146 barrels hourly from a pay horizon running from 2,656 to 2,710 feet.

Other companies wanted to participate in the rich Van field but decided to forego the usual wastefulness associated with competitive drilling. The Humble, Texas, Sun, and Shell companies entered into an agreement with Pure, the major lessee in the field (with over 80 percent of the acreage) to develop the field jointly, or as a unit, with each receiving a percentage equal to its investment. Lawyers Hines H. Baker of Humble and Robert A. Shepherd of Pure drafted the agreement.

This unitization pointed the way to sound conservation practices, for it avoided the dissipation of the field's resources and eliminated needless duplication of drilling expenses. Of course, such restrictions also smacked of restraint of trade, and the parties to the agreement risked antitrust prosecution, but none resulted. This potential problem with unitization of the Van field served as a forewarning of serious troubles that would shortly develop in the East Texas field. Small operators there saw prorationing, a different way to restrict production and therefore promote conservation, as a tool of the major companies to drive them from the field.

Unitization of the Van field meant that its development differed sharply from those that had gone before. It was sober and orderly—no vice, lawlessness, or social chaos. With the big companies running the entire show, there was little opportunity for the competitive scramble that usually attracted criminal elements to an oil field. Symbolic of the staid nature of the Van field, the First Methodist Church sat right in its middle, with half a dozen wells producing on its property. Other benefits of unitization were that the developers drilled few dry holes (out of a total of 600 wells) and that underground gas pressure was maintained so that oil continued to flow freely. Even with production restricted to maximize the field's life, Van had yielded over 15 million barrels by 1932.

The seismograph recorded waves from explosives set off in the earth. The nature of the waves indicated whether oil deposits were likely. The Van field was discovered by a seismograph crew of the Geophysical Research Corporation in June, 1927, although production didn't begin for another two years. *Smithsonian Institution.*

On October 14, 1929, the Van field proved a rich discovery, drawing further attention to the possibilities of East Texas. At the end of the month the stock market crashed, throwing the petroleum industry, as well as all other phases of American business, into a tailspin. Those who dreamed of making a fortune from Van, twenty miles northwest of Tyler, had to modify their expectations. *Shell Oil Company.*

This 122-foot derrick in the East Texas field was probably constructed from yellow pine cut near the well. Carpenters building derricks for wildcatters often alternated between working for wages and for an interest in the well. *Houston Museum of Natural Science.*

# East Texas

SOMEWHAT ironically, discovery of the world's greatest oil field coincided with the Great Depression. Abnormal times prevented those participating in the fruits of the discovery from realizing normal monetary returns. Then the great glut of oil from the East Texas field further depressed the price of crude oil and the disappointment deepened. Yet, the oil boom brought jobs, albeit low-paying, to many who would otherwise have been jobless, and the oil resulted in some income, however depreciated, that was better than nothing. Whatever else the field did, it relieved the depression tedium for many East Texans.

The hero of the East Texas field was Columbus M. ("Dad") Joiner, a grizzled wildcatter from Ardmore, Oklahoma. Although professionals had pronounced the area devoid of petroleum, Joiner had built his entire career on playing his hunches. He had made and lost two oil fortunes in Oklahoma and came to East Texas somewhat down on his luck. He scraped together enough money to buy leases on land in Rusk County between Henderson and Overton. The hilly land was not prosperous, for the red clay supported only marginal farming and grazing. In 1927 Joiner found a geologist to work with him, the rotund A. D. ("Doc") Lloyd of Fort Worth. Lloyd studied the 5,000 acres under lease, mapping the underground structure. The most likely spot for drilling lay on Mrs. Daisy Bradford's farm.

Before drilling could begin, Joiner had to raise some cash. He did this by peddling shares in his project in Dallas and other cities. Virtually every aspect of his drilling operation was financed piecemeal, and often work had to stop until Joiner secured a few more dollars. With the operation underfinanced from the beginning, the drilling crew never had adequate equipment. The drilling rig, boilers, pipe, and bits were veterans of countless other explorations. With a little cash in hand, Joiner hired Ed C. Laster as head driller and the requisite crew. Work began in August, 1927, but because of the antiquated equipment, it proceeded slowly. A drill pipe stuck at 1,098 feet and could not be fished out, so in February, 1928, the crew abandoned the hole. Joiner began another well, drilled deeper, but still found nothing.

These failures did not shake Joiner's resolution to drill a third time on the Bradford property, despite his being seventy-three, in ill health, and broke. He sold more stock in the venture, raising enough money to spud in Daisy Bradford #3, about two miles southeast of his initial well. During much of the drilling, Joiner had to be elsewhere, cajoling and pleading for the cash to continue. Money was so scarce that the fireman had to burn scrub pine in the boiler. Around the first of September, 1930, Laster found traces of oil from the Woodbine sand and summoned Joiner from Dallas. News of the well's potential traveled quickly, especially through the network of oil scouts. All major companies employed scouts to report regularly on exploration throughout the state. Like minstrels of old, they kept crews at isolated locations informed of what was going on elsewhere. Sometimes they were viewed with suspicion, and sometimes drillers tried to dupe them by "salting" a core sample with crude from another location, but scouts were an accepted part of the oil fraternity. A scout from a major company looked at Bradford #3 but concluded that Laster had salted the sample and did nothing. The local folk were not so jaundiced and flocked to the well to look at the oil-soaked cuttings.

On September 5, the road to Joiner's well was packed with Model T's and Model A's, as one thousand curious people watched the drilling. Those present at 8:30 P.M. witnessed the opening of the great field. Daisy Bradford #3 erupted from 3,592 feet, gushing over the crown block 112 feet above the derrick floor. After a few minutes, the crew closed the valve and mudded in the well to prevent a blowout. News of Joiner's well electrified East Texas, with lease buyers swarming over the countryside. A month earlier, leases had gone begging at $10 an acre, but now they sold for $500 to $1,000. Henderson, seven miles from the well and the largest town nearby, was jammed with visitors who strained all the guest facilities. The contagion spread quickly to Overton, Kilgore, Tyler, Longview, and Gladewater. Roads throughout the area were clogged with cars and people.

Other derricks were quickly erected near Joiner's well, and a community developed nearby. Appropriately, it became known as Joinerville. As quickly as lumber could be imported, the components of a town began to take shape: grocery stores, rooming houses, cafes, and makeshift dwellings. Today all that remains of Joinerville is a post office operating in a little grocery story. Down the road near the discovery well, the state has erected a monument to Joe Roughneck.

Dad Joiner's monument is the entire East Texas field. As other wells were developed in the area, people began to realize with astonishment that it was all one giant field, forty-three miles long and five miles wide. Joiner had managed to tap the edge of the world's largest pool of oil, most of which lay to the west and north of Bradford #3. If he had drilled just one-quarter mile further east, he would have had another dry hole and most likely would have given up in failure and utter despair. The landmark wells in the East Texas field were Joiner's; the Lou Della Crim #1, near Kilgore and a dozen miles north of Joiner's; and Lathrop #1 between Gladewater and Longview, another fifteen miles north of Kilgore.

While Joiner rejoiced that his faith in the field had been justified, he had mixed feelings about what to do with the property once it began producing. He had seen other fields wax and wane, so he decided to hedge his bets by selling part of his East Texas interests. At the end of November he sold 4,000 acres of leases to H. L. Hunt, who had already grown rich in the oil business at El Dorado, Arkansas. Hunt paid Joiner $1.25 million for those leases which formed the cornerstone for Hunt's tremendous expansion in the Texas oil business.

On December 27, 1930, the second landmark well of the field, the Lou Della Crim #1, startled Kilgore when its great gas pressure began hoisting a potential of 22,000 barrels daily into the sky. The drilling contractor, William E. Cain, brought the well under control quickly. The lease was owned by Ed W. Bateman of Fort Worth, and the Crim family from whom he bought it was prominent in the affairs of the 1,000-resident town. Like Henderson, Kilgore quickly filled with fortune seekers, land men, and scouts, and its facilities were strained to the uttermost. It began frenzied construction to take care of the 10,000 people it expected within a few months.

Because of the splendid showing of the Kilgore discovery well, Humble bought the Crim lease for $2.1 million. It began drilling at 150-foot intervals along the border of its lease. Since the field was three and one-half miles from town, it became a major problem hauling drilling equipment over the rough terrain, especially in rainy weather, which East Texas featured.

Longview eagerly hoped to participate in the East Texas field, its chamber of commerce offering $10,000 for the first well in its environs. That award went to the drillers of the F. K. Lathrop #1 in February, 1931. The well was located six miles northwest of Longview. This discovery created the same excitement and congestion in Longview as the earlier wells had in Henderson and Kilgore. Its population doubled to 10,000 within two months. Among those who came were representatives of major companies, Humble, Magnolia, and Gulf. All had offices in Longview to supervise their large holdings in the field.

By the time the dimensions of the East Texas field were fully established, it covered parts of five counties: Rusk, Gregg, Upshur, Smith, and Cherokee, with the greatest concentration in the first two. Dad Joiner's well led to such rapid drilling that sixteen months later 3,732 wells had been completed. By the end of 1932, the field boasted 5,652 producing wells. The problem was they produced crude oil for which little demand existed. At the time of Daisy Bradford #3, crude sold for $1.10 a barrel, but with the glut of East Texas oil in the summer of 1931, it brought only a nickel.

The problem was clearly overproduction, combined with the faltering depression economy. A simple solution would be to restrict production to make the supply more equal to the demand so that prices would rise, but it really was not that simple. For scores of small operators in the East Texas field, even a nickel or dime a barrel was better than the nothing they would have otherwise, so they resisted various attempts at restriction or prorationing.

Suggestions for prorationing came first in February, 1931. Owners of leases, land, and royalties and some small producers met in Tyler to discuss the benefits of restricting the output of oil. From the meeting a permanent organization resulted, the East Texas Lease, Royalty, and Producers Association. Another committee, representing only Rusk and Gregg counties, urged the Texas Railroad Commission to initiate proration hearings quickly.

When the East Texas Lease, Royalty, and Producers Association met, its vice-president, W. B. Hamilton, predicted, "If you permit just anyone to take your market outlet, you'll have anarchy among operators; you will witness men taking the law into their own hands." Despite the manifest logic of this position, others, including Doc Lloyd, thought prorationing too restrictive. A sufficient number of small landowners, leaseholders, and operators agreed with Lloyd to prevent any consensus for prorationing in the East Texas field. Thus, without any voluntary self-regulation, the matter was left in the hands of the railroad commission, which was empowered by law to act.

Its first pronouncement came in April, 1931, after hearing testimony from interested parties. Initially, the field could produce 70,000 barrels daily, to be increased to 100,000 as more wells were completed. The commission hoped, of course, that drilling would cease so that it could deal with a known flow of crude. Unfortunately, the commission's quotas were always behind the field's pace. Since the field produced 140,000 barrels daily when the 70,000 limit was imposed, the commission upped its figure to 90,000, plus an increment of 90,000 over a three-month period. The week prior to the commission's revised ruling, the field soared to 340,000 barrels a day. There seemed to be no way that the commission's figures related to the reality of the field. Nor did that matter, for operators blithely ignored the commission's orders. They made it plain that they intended to run as much oil as possible. That produced over the quota was called "hot oil."

The commission's timidity was entirely justified. It had little backing by other elements of the state government. Pressure from small producers resulted in the legislature's frequently changing the laws under which the commission operated. These producers also filed suits to enjoin the commission from issuing quotas, and the courts succeeded in undoing its efforts. The Macmillan case most emphatically undercut the commission's authority.

Property owners in the East Texas field, the Macmillans sued to prevent the commission from restricting production on their property. They claimed that the commission's orders resulted in price-fixing rather than the prevention of physical waste of oil. A three-judge district court in West Texas heard the case on June 24, 1931, and held for the Macmillans the following month. Its decision incorporated the ideas that the commission regulations did not regulate physical waste, but that they related to market demand and therefore resulted in fixing prices to prevent economic waste. A state statute specified that waste did not include economic waste. The commission could act legally only to prevent physical waste, so its prorationing activity was declared illegal.

The court's ruling had an immediate effect on Governor Ross S. Sterling. Although he had earlier summoned the legislature to strengthen the commission's hand, he took his cue from the court and announced that he would veto any legislation that sought to regulate the flow of East Texas oil on the basis of demand, because of its price-fixing implications. As a former president of Humble, Sterling must have winced over this decision. The major companies had long favored orderly production, since that kept supply equal to demand and prices up. That orderly production also conserved natural resources indicated that profits could be expected over the long run. The wanton exploitation of the East Texas field meant that gas pressure would soon be dissipated and the life of the field foreshortened, with resultant losses for everyone. With this action as governor, Sterling violated what he knew was in the best interests of the industry and ultimately of the people.

Because of the explicitness of the court and governor on the matter of economic waste, the legislature passed a bill on August 12, 1931, empowering the commission to regulate to avoid physical waste. The conservation forces had finally found an advocate! Three days after passage of the bill, Sterling ordered all wells to shut down in Rusk, Gregg, Smith, and Upshur counties. He dispatched the Texas National Guard, commanded by General Jacob F. Wolters, to the field to maintain order. Threats had flown back and forth from those who supported and opposed the law that they would dynamite each others' storage tanks and ignite wells. By mid-August the field was producing a staggering 848,398 barrels daily.

That commandant of the National Guard was Jacob F. Wolters did not escape notice. In civilian life he was the chief counsel of the Texas Company. Those opposing prorationing saw his leading troops in their field as evidence of the oppression of small producers by the majors.

Guard troops set up camp on "Proration Hill" outside Kilgore and took charge of the field. They ensured that commission quotas were followed. The allowable for each well on September 18 was 185 barrels; the next month it dropped to 165. In neighboring Oklahoma Governor William H. ("Alfalfa Bill") Murray had already established the precedent of the state government's closing down its oil fields. He imposed martial law, saying troops would not be withdrawn until Oklahoma crude sold for one dollar a barrel.

Enforced prorationing in East Texas had some desired effect. Not only was oil conserved but the state earned $1.6 million in production taxes. When troops occupied the field, oil sold for twenty-four cents a barrel. In mid-March 1932 it had climbed to sixty-seven cents and by early November it hit eighty-two cents. These figures convinced most operators of the soundness of regulation.

But not all. Disgruntled operators filed suit to protest armed forces in the field. A federal court decided on December 12, 1932, that only an insurrection or the direct threat of one justified martial law. In the Constantin case, the court declared the presence of troops illegal. Sterling fudged on the implications of the court's decision, deciding to leave the National Guard in the field as peace officers but not to enforce prorationing. Even when they were there to ensure that wells produced only their allowables, the guardsmen encountered stiff opposition and deception.

Hot oil operators had a full bag of tricks with which to evade the law. Threats of violence and bribery were common. One inspector, when finally caught, admitted that in one month he had taken $2,000 in bribes. Ingenious operators employed many mechanical devices to run hot oil. They dug several dummy wells next to a producer, supplying them with pipes through which oil from the real well could flow. With each well having its allowable, that was multiplied by the number of dummy wells. Dishonest operators would tap pipelines or tanks, routing the oil to "moonshine" refineries or loading it in trucks. Bypass pipelines were popular, but they presented the problem of where to locate the valves. A resourceful crook put a valve behind his bathtub. Another devised a left-handed valve that appeared to be off when it was fully open. Sometimes oil was smuggled out of the field in gasoline trucks disguised as moving vans. Fines proved insufficient deterrent to breaking the law. If a man could earn $8,000 a day running hot oil, he could afford a $1,000 fine.

The legislature acted further to strengthen the power of the railroad commission. In November, 1932, it passed a market demand law that geared prorationing to the market so that producers would receive a fair price. Inspection of refineries was made easier when the national Congress taxed all oil produced in the country. Refineries had to document where their crude had originated. When Congress passed the Connally "hot oil" bill early in 1935, oil-field regulation throughout the country was largely in federal hands. It had been sponsored, fittingly, by Texas Senator Tom Connally.

The economic and physical waste borne by the East Texas field up to that time was staggering. The field had 17,650 wells with over 700 derricks standing within Kilgore's city limits. Experts estimated that half that number of wells could have produced as much oil, which had found an inadequate market. They reckoned that the waste in drilling costs ran from $100 million to $150 million. With waste rampant, East Texas operators could take scant comfort in the fact that their field turned out to be the world's biggest.

As America's Great Depression merged into its preparation for World War II, the early phase of Texas' oil history ended. From the beginning of the industry in Corsicana through the development of the East Texas field, quick profits had been the motive. The rule of capture reigned supreme, enabling many to amass wealth quickly but at the expense of greater returns over a longer period. Social utility was a concept that most wildcatters never dreamed of.

Those who participated in Texas oil in its early days left an impressive record. They created the state's largest industry, producing billions of dollars of wealth. In the process, they transformed the state's economy from an agricultural to industrial basis. Both individual entrepreneurs and corporate giants contributed to the development of the industry, the former a bit more colorful than the others. The personnel in the business differed widely—from the roustabout on a lonely rig in the Yates field to the production accountant, dressed in coat and tie, sitting in his Houston office. By the 1930's the major oil companies easily dominated the industry, with their integrated operations ranging from the geologist's investigations through the pump attendant filling the motorist's tank.

Even so, the clearest symbol of the early days of Texas oil emerged at the end of that era: the individual wildcatter, working for himself. Dad Joiner was a wildcatter, the best of the breed. The majors had had their chance with the East Texas field but had decided nothing was there. Playing his hunches, with insufficient money to do the job properly, Joiner discovered the world's largest pool of oil. His success proved that individual initiative and insight still had their rewards, even in corporate America. Thus, Joiner, the lone prospector, gambling with his own and what additional money he could scrape up, symbolizes best the early Texas oil industry.

The prelude to the giant East Texas field came with the Colliton discovery well in Cherokee County, November 9, 1924. J. A. Colliton in the business suit. *Southwest Collection*.

East Texas / 229

C. M. ("Dad") Joiner, a veteran wildcatter, tirelessly promoted drilling in Rusk County during the late 1920's. Despite technical and financial difficulties, he persevered with his third well on Daisy Bradford's farm, about halfway between Henderson and Overton. An expectant throng flocked to the well site on September 5, 1930, and witnessed the discovery of the world's greatest oil field. *Texas Mid-Continent Oil & Gas Association*.

Joiner and his corpulent geologist, A. D. ("Doc") Lloyd, shaking hands at the Daisy Bradford #3. *Texas Mid-Continent Oil & Gas Association*.

*Left:* The crew on this rotary rig in the East Texas field has lowered the crown block to attach some equipment. *Houston Museum of Natural Science. Right:* High on the derrick to rig the well, this worker had a good view of the East Texas terrain. *Kilgore News Herald.*

Before the boom came to Kilgore, the boys gathered at Brown's Drug Store to enjoy a Coke. *Kilgore News Herald.*

*Left:* The J. A. Knowles #1, first well in the town of Kilgore, which was transformed by the boom. *Texas Mid-Continent Oil & Gas Association. Right:* Oil brought prosperity to the Crim family of Kilgore. On the outskirts of town, the Lou Della Crim #1, a dozen miles north of Joiner's discovery well, proved that the East Texas field was huge. Kilgore's craze for oil led to intensive drilling all over town. Here a collapsed derrick damaged Mayor Crim's house. *American Petroleum Institute.*

The best thing that could be said about these 1931 Kilgore residences was that they were close to the job. *Southwest Collection.*

While mud plagued all oil towns, more frequent rain in East Texas produced more acute problems, as on Moody Street in Kilgore. *American Petroleum Institute*.

The Kilgore depot and traffic along Commerce Street. The pipes in the foreground signify an oil field. *Kilgore News Herald*.

*Left:* Ross S. Sterling as candidate for governor in 1930. As governor, he used the Texas National Guard to enforce prorationing in the East Texas field. *UT Archives.* *Right:* The Texas National Guard (the 36th Infantry Division) was commanded by General Jacob F. Wolters (*left*), who as a civilian was chief counsel for the Texas Company. Independents understood his willingness to enforce the governor's orders. *Exxon Corporation.*

Governor Ross Sterling ordered the East Texas field shut down on August 15, 1931, to prevent wastage of oil. Later proration quotas allowed wells to produce only a fraction of their capacity. A crowd gathered in Kilgore to protest Sterling's ruling, which he had enforced with the National Guard. The plain people saw prorationing as a tool of the majors to deny independent producers the chance to earn a living. *Kilgore News Herald.*

The Texas National Guard moving into the East Texas field to halt illegal production of oil. *Kilgore News Herald.*

This Gladewater photographer seemed to sympathize with the independents. *Texas Mid-Continent Oil & Gas Association.*

Once prorationing was relaxed, flush production resumed, as at this East Texas well. *Southwest Collection*.

Kilgore in the mid-1930's. *American Petroleum Institute*.

In early February, 1931, when the F. K. Lathrop #1 came in six miles northwest of Longview (and fifteen miles north of Lou Della Crim #1), the vast extent of the East Texas field was defined. The Longview chamber of commerce awarded $10,000 to the drillers, Jeffries and Foster. *Exxon Corporation*.

The Moriah Cemetery in the East Texas field. *Texas Mid-Continent Oil & Gas Association.*

The Locke Refinery in Gladewater, fifteen miles west of Longview. Such small operations typified the East Texas field. *UT Archives.*

Humble's truck fleet at Gladewater. *Exxon Corporation.*

Humble used this 1934 truck in the East Texas field to measure the amount of natural gasoline contained in each thousand feet of casing head gas from a well. Knowing how many gallons of natural gasoline were produced in each thousand cubic feet of casing head gas enabled gas processing plants to credit the proper amount of natural gasoline to each lease. In turn, payments to each lease were determined. *Exxon Corporation.*

Transportation of oil and gas is less identified with place than production and refining. Here an early pipeline tong crew willingly rests for the photographer. With their tongs, workers screwed together links of pipe, which sometimes leaked at the joints. Later, welding joints proved effective in stopping leaks. *American Petroleum Institute.*

# Epilog

THE petroleum industry conventionally divides into four components: production, refining, transportation, and marketing. The first two have sometimes occurred in close proximity, for obvious economic reasons. But when production was far from any population center that would furnish purchasers of refined products, crude oil had to be transported to refineries by rail, water, or pipeline. Because production and refining in Texas have often been close to each other, this book has treated them together in its geographic organization. But not all elements of Texas oil fit into the historian's neat categories, just as life itself usually defies tidy labels. Consequently, the book includes some photographs that document the geographical diversity of the state's experience with oil, as well as the transportation and marketing aspects of the industry.

Because of the expectation of making a profit from oil wells, wildcatters searched throughout the state. Frequently they discovered oil in small quantities—not enough to create a boom, but enough to return a modest profit. Thus, Marlin had a little oil flurry in 1910, and a few wells went down in the South Bosque field, southwest of Waco, in the 1920's. In many places, production was insufficient for workers to form a town, as in the South Bosque field and in the wells in Throckmorton and Williamson counties. The photographs of these last two offer interesting, if different, aspects of Texas oil. In Throckmorton County oil production blended harmoniously with an older economic enterprise—ranching. Such coexistence was widespread, even if some of the old wranglers did not appreciate the intrusion. What they did tolerate was the oil income. The oil field in Williamson County was near Taylor—as close as oil got to Austin, the state's capital. Whereas the governor of Oklahoma could look out his office window in 1930 and see producing wells on the capitol grounds, the Texas governor would have had to journey to adjoining Williamson County for similar inspiration. In such diverse places as Olney, Rising Star, and Freeport, oil derricks stood as lonely sentinels over the countryside. At Freeport oil ran a poor second to sulphur as a natural resource.

While the East Texas field was developing into its full dimensions, one other important field attracted attention, Conroe. Discovered during the depths of the Great Depression, the field did not reward its developers as it would have in better times, but it did demonstrate that oil could still be extracted profitably from the vast expanse of the coastal plain. Further drilling at nearby Tomball, as well as other locations, proved the point. In the Conroe field drillers made a significant innovation in 1934. They showed that burning wells could be brought under control by directional drilling. To shut off the fire's supply of crude, a crew some distance from the flaming well would drill at an angle to intersect the flow of oil. It could then be either diverted or capped.

In the oil business, transportation has been an integral part of the profit equation. Before oil could be sold, it had to be refined, and for refining, it had to be transported from the place of production. Although this distance could be quite short, as from the Goose Creek field to the Baytown refinery or in the East Texas field (see photograph on page 238), it usually amounted to many miles. Both railroad tank cars and sea-going tankers carried crude and refined oil. Understandably, photographers took pictures of these means of transportation when the trains or ships waited at terminals, so those photographs are usually identified with a specific place. Pipelines, on the other hand, rarely were photographed near a terminal. Photographers evidently thought that the wide open spaces offered the typical environment for a pipeline picture, and appropriately so. The industry used pipelines when oil had to be transported economically and reliably over great distances, as from the Big Lake field to coastal refineries. The only time pipeline pictures showed any action or human involvement was during their construction, so the open trench predominates in such photographs.

Marketing constitutes the final phase of the oil business—where consumers' payments determine the ultimate profitability of the other phases of the industry. Here competition among the majors and between the major and minor companies or independents was keen. The competition was in the areas of advertising and service, as well as in price. As the photographs show, Magnolia, Humble, Texaco, and Gulf strove to present the image of efficient and courteous delivery services, as well as attractive filling stations. What photographers did not capture was the marketing competition between the major companies and the small independents. The gas wars pitting the majors against the independents during the Great Depression escaped photographic documentation, as did the modest, if not primitive, marketing facilities of the latter. (Such photographs, if ever taken, are not in research collections.)

Although this photographic record of Texas oil began in 1866, that early date is somewhat misleading because Barret's well at Oil Springs did not launch a stable, continuing industry. That honor must be reserved for the 1896 Corsicana well. In essence, then, the early phase of the Texas oil industry covered forty years, until 1936. Any cut-off date for this phase is manifestly arbitrary, but the Texas Centennial year seems logical enough. The wildcatter had proved his enduring worth in discovering the East Texas field, whose development continued until the middle of the 1930's. The latest photograph in the book (page 50) dates from 1936 and shows one of Glenn McCarthy's wells that proved a total loss. McCarthy, the flamboyant wildcatter whom Edna Ferber used as a model for Jett Rink in *Giant*, went on to earn and lose a sizable fortune, not uncommon for the breed.

With the coming of World War II and its voracious demands for petroleum products, the nature of the Texas oil industry changed appreciably. Corporate control became far more evident during the war and afterward. As the major companies developed new fields, especially in the Permian Basin, orderliness replaced the rough-and-tumble ways of a Batson or Desdemona or Burkburnett. The early days of Texas oil had passed.

This 1910 well in Marlin boasted the first 120-foot derrick in Texas. The crew (*left to right*), Fred Jeffries, Si Young, Bill Allen, and M. A. Johnson, used the first string of tool joints manufactured by the American Well and Prospecting Company. *UT Archives.*

Drilling Amicable #3 at South Bosque, a small field southwest of Waco in McLennan County. *Humanities Research Center, UT.*

Humble tank truck on a Waco street in the 1920's. The middle unit was more humble than flashlike. *Exxon Corporation.*

*Left:* Cattle fatten best by feeding on plains enriched by oil, as on this Throckmorton County ranch. *Southwest Collection. Right:* Swastika #1 gushing near Olney, about forty-five miles south of Wichita Falls. In the 1920's these Young County folk had probably never heard of Hitler, who would desecrate the ancient good luck symbol. *Southwest Collection.*

A 1930 view of a Williamson County field near Taylor, northeast of Austin. *Library of Congress.*

Heavy oil-field equipment clogs the Olney street in the late 1920's. Trucks belonged to H. B. Lambert. *Southwest Collection.*

*Left:* Three wells at Rising Star, twenty-seven miles southwest of Desdemona in Eastland County, 1930. If the two on the right went dry, the farmer prayed that the one on the left didn't. *Exxon Corporation. Right:* Nature adorned this derrick at Freeport, on the Gulf coast. *Houston Public Library.*

*Left*: Strake #2 at Conroe, completed June 5, 1932, at 5,026 feet. It produced 5,000 barrels daily. George W. Strake, who discovered the field, spudded in his first well with a shovel, lacking time to get proper equipment onto the well site before the lease expired. Strake was at mass on December 13, 1931, when his first well blew in and he rebuffed the crew for disturbing him with the news while he worshipped. *Houston Museum of Natural Science. Right:* Drilling the "Rathole" at Tomball, 1934. *Houston Museum of Natural Science.*

The crew that drilled the Mrs. H. N. Moore #16 at Conroe in February, 1933. They put down a 5,100-foot well in seventeen days, a record for the time. *Texas Mid-Continent Oil & Gas Association.*

Epilog / 247

Digging a pipeline trench through this rocky terrain was no cinch, so the laborers gladly paused to have their picture taken. *American Petroleum Institute.*

A tong crew lays an early pipeline. *American Petroleum Institute.*

An ever-present danger in pipeline work is a cave-in. Here an injured worker is carried out of the ditch. *American Petroleum Institute.*

This 1930 ditch-digging machine eliminated some of the onerous physical labor connected with laying pipelines. *American Petroleum Institute.*

An open-air cooler at a natural gas pumping station which moved the gas through the pipes to the point of use. *Smithsonian Institution.*

A 1915 delivery of oil in Normangee, eight miles above North Zulch. *American Petroleum Institute.*

Delivery men for San Antonio's Producers Refining Company prepare to transport their products to markets. *Humanities Research Center, UT.*

Judging from the starched collar, the man behind the wheel of 592 wasn't the regular driver. He probably worked in Magnolia's San Antonio office. *Humanities Research Center, UT.*

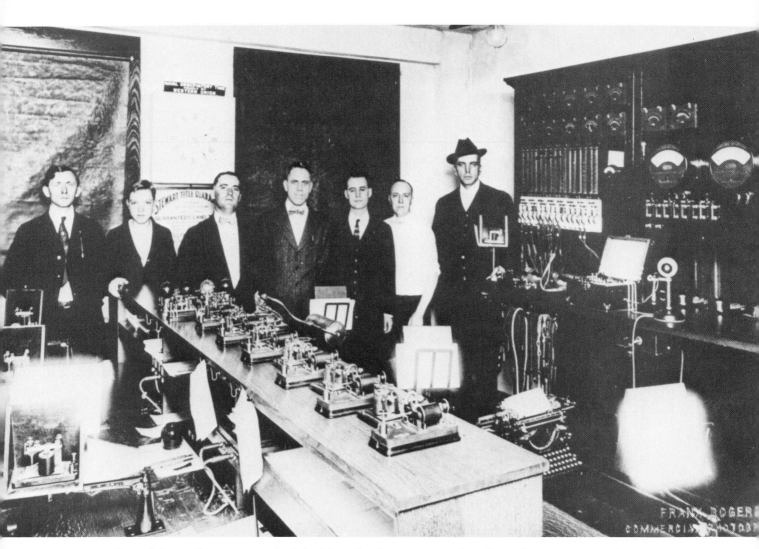

Part of the marketing network of the Magnolia Company was the telegraph room in its Dallas office, 1923. For years the flying red horse atop the Magnolia Building dominated Dallas' skyline. *American Petroleum Institute.*

In the early 1920's this tank wagon delivers oil for the basement furnace. *American Petroleum Institute*.

"Trust your car to the man who wears the star." *Exxon Corporation*.

Gulf offered the latest in filling stations and tank trucks in the early 1930's. *Gulf Oil Co.*

Like other oil companies, Humble had standard designs for filling stations. In the 1930's, this design could be found anywhere from Jefferson to Marfa. *Exxon Corporation.*

# Bibliography

Boatright, Mody C., and William A. Owens. *Tales from the Derrick Floor: A People's History of the Oil Industry*. Garden City, N.Y.: Doubleday & Company, Inc., 1970. Consists largely of interviews with Texas oil pioneers.

Gard, Wayne. *The First 100 Years of Texas Oil & Gas*. Dallas: Texas Mid-Continent Oil & Gas Association, [1966]. A pamphlet containing most of the basic facts about production.

Hagy, Lawrence. "Borger—The Last of the Oil Field Booms." Manuscript of speech before the Desk and Derrick Club of Amarillo, December 5, 1972.

King, John O. *Joseph Stephen Cullinan: A Study of Leadership in the Texas Petroleum Industry, 1897–1937*. Nashville: Vanderbilt University Press, 1970. Demonstrates Cullinan's major role in developing the Texas oil industry.

Landry, Wanda A. "Lawlessness in the Big Thicket." Master's thesis, Lamar University, 1971. Treats crime in the Saratoga and Batson oil fields.

Larson, Henrietta M., and Kenneth Wiggins Porter. *History of Humble Oil & Refining Company: A Study in Industrial Growth*. New York: Harper & Brothers, 1959. The company's official history, detailed and encyclopedic.

Moore, Richard R. *West Texas after the Discovery of Oil*. Austin: Jenkins Publishing Co., 1971. Shows the ways that oil affected the life and economy of the region.

Rister, Carl Coke. *Oil! Titan of the Southwest*. Norman: University of Oklahoma Press, 1949. An invaluable source of information on the industry in Texas as well as Kansas, Oklahoma, Louisiana, and New Mexico.

Rundell, Walter, Jr. "Centennial Bibliography: Annotated Selections on the History of the Petroleum Industry in the United States." *Business History Review* 33 (Autumn, 1959): 429–447.

———. "Texas Petroleum History: A Selective Annotated Bibliography." *Southwestern Historical Quarterly* 57 (October, 1963): 267–278.

Tolbert, Frank X. *The Story of Lyne Taliaferro (Tol) Barret, Who Drilled Texas' First Oil Well*. Dallas: Texas Mid-Continent Oil & Gas Association, 1966. A thoroughly researched pamphlet.

Trevey, Marilyn D. "The Social and Economic Impact of the Spindletop Oil Boom on Beaumont in 1901." Master's thesis, Lamar University, 1974.

Warner, C. A. *Texas Oil and Gas since 1543*. Houston: Gulf Publishing Company, 1939. The foundation upon which all subsequent studies of the Texas oil business have been based. Encyclopedic data.

## Suggestions for Further Reading

Beaton, Kendall. *Enterprise in Oil: A History of Shell in the United States*. New York: Appleton-Century-Crofts, 1957.

Boatright, Mody C. *Folklore of the Oil Industry*. Dallas: Southern Methodist University Press, 1963.

———. *Gib Morgan, Minstrel of the Oil Fields*. Austin: Texas Folklore Society, 1945.

Clark, James A., and Michel T. Halbouty. *The Last Boom*. New York: Random House, 1972. The East Texas field.

———. *Spindletop*. New York: Random House, 1952.

James, Marquis. *The Texaco Story: The First Fifty Years, 1902–1952*. New York: The Texas Company, 1953.

King, John O. *The Early History of the Houston Oil Company of Texas, 1901–1908*. Houston: Texas Gulf Coast Historical Association, 1959.

Knowles, Ruth Sheldon. *The Greatest Gamblers: The Epic of American Oil Exploration*. New York: McGraw-Hill Book Company, 1959.

Leeston, Alfred M.; John A. Crichton; and John C. Jacobs. *The Dynamic Natural Gas Industry*. Norman: University of Oklahoma Press, 1963.

McDaniel, Ruel. *Some Ran Hot*. Dallas: Regional Press, 1939.

Myres, Samuel D. *The Permian Basin, Petroleum Empire of the Southwest: Era of Discovery, from the Beginning to the Depression*. El Paso: Permian Press, 1973.

Owens, William A. *Fever in the Earth*. New York: G. P. Putnam's Sons, 1958. The best novel of the oil industry, set at Spindletop.

Schwettmann, Martin W. *Santa Rita: The University of Texas Oil Discovery*. Austin: The Texas State Historical Association, 1943.

Spratt, John S. *The Road to Spindletop: Economic Change in Texas, 1875–1901*. Dallas: Southern Methodist University Press, 1955.

Tait, Samuel W., Jr. *The Wildcatters: An Informal History of Oil-Hunting in America*. Princeton: Princeton University Press, 1946.

Thompson, Craig. *Since Spindletop: A Human Story of Gulf's First Half Century*. Pittsburgh: Gulf Oil Corporation, 1951.

Williamson, Harold F., and Arnold R. Daum. *The American Petroleum Industry: The Age of Illumination, 1859–1899*. Evanston: Northwestern University Press, 1959.

———; Ralph L. Andreano; Arnold R. Daum; and Gilbert C. Klose. *The American Petroleum Industry: The Age of Energy, 1899–1959*. Evanston: Northwestern University Press, 1963.

# Index